Machine Learning and Deep Learning for Smart Agriculture and Applications

Mohamamd Farukh Hashmi
National Institute of Technology, Warangal, India

Avinash G. Kesakr
Visvesvaraya National Institute of Technology, India

A volume in the Advances in Environmental Engineering and Green Technologies (AEEGT) Book Series

IGI Global
PUBLISHER of TIMELY KNOWLEDGE

Published in the United States of America by
IGI Global
Engineering Science Reference (an imprint of IGI Global)
701 E. Chocolate Avenue
Hershey PA, USA 17033
Tel: 717-533-8845
Fax: 717-533-8661
E-mail: cust@igi-global.com
Web site: http://www.igi-global.com

Library of Congress Cataloging-in-Publication Data

Names: Hashmi, Mohammad Farukh, editor. | Keskar, Avinash G., editor.
Title: Machine learning and deep learning for smart agriculture and applications / edited by Mohammad Farukh Hashmi and Avinash G. Keskar
Description: Hershey, PA : Engineering Science Reference, [2024] | Includes bibliographical references and index. | Summary: "As health monitoring of plants and fruits is essential for sustainable agriculture. In the agriculture farming business, plant diseases are the major reason for monetary misfortunes around the globe. Automated detection and identification of plant diseases using sophisticated image processing coupled with advanced computer vision techniques results in accurate and fast identification with less human effort and labor cost. This book is to to provide an efficient solution for the detection of plants and fruit diseases, using various machine learning and deep learning techniques. The objective of the book is to find out how agriculture is involved with prepared and efficient profound learning algorithms. There are huge prospects for agriculture administration through the proper use of Machine Learning, Deep Learning, Big Data, and IoT structures. The majority of agricultural research institutes have already adopted Internet of Things for the monitoring of a variety of farm services"-- Provided by publisher.
Identifiers: LCCN 2023029630 (print) | LCCN 2023029631 (ebook) | ISBN 9781668499757 (hardcover) | ISBN 9781668499764 (ebook)
Subjects: LCSH: Agricultural innovations. | Machine learning. | Deep learning (Machine learning).
Classification: LCC S494.5.I5 M323 2024 (print) | LCC S494.5.I5 (ebook) | DDC 338.1/6--dc23/eng/20230726
LC record available at https://lccn.loc.gov/2023029630
LC ebook record available at https://lccn.loc.gov/2023029631

This book is published in the IGI Global book series Advances in Environmental Engineering and Green Technologies (AEEGT) (ISSN: 2326-9162; eISSN: 2326-9170)

British Cataloguing in Publication Data
A Cataloguing in Publication record for this book is available from the British Library.

Advances in Environmental Engineering and Green Technologies (AEEGT) Book Series

ISSN:2326-9162
EISSN:2326-9170

Editor-in-Chief: Sang-Bing Tsai, Zhongshan Institute, University of Electronic Science and Technology of China, China

MISSION

Growing awareness and an increased focus on environmental issues such as climate change, energy use, and loss of non-renewable resources have brought about a greater need for research that provides potential solutions to these problems. Research in environmental science and engineering continues to play a vital role in uncovering new opportunities for a "green" future.

The **Advances in Environmental Engineering and Green Technologies (AEEGT)** book series is a mouthpiece for research in all aspects of environmental science, earth science, and green initiatives. This series supports the ongoing research in this field through publishing books that discuss topics within environmental engineering or that deal with the interdisciplinary field of green technologies.

COVERAGE

- Green Transportation
- Sustainable Communities
- Green Technology
- Electric Vehicles
- Biofilters and Biofiltration
- Alternative Power Sources
- Water Supply and Treatment
- Waste Management
- Contaminated Site Remediation
- Cleantech

The Advances in Environmental Engineering and Green Technologies (AEEGT) Book Series (ISSN 2326-9162) is published by IGI Global, 701 E. Chocolate Avenue, Hershey, PA 17033-1240, USA, www.igi-global.com. This series is composed of titles available for purchase individually; each title is edited to be contextually exclusive from any other title within the series. For pricing and ordering information please visit http://www.igi-global.com/book-series/advances-environmental-engineering-green-technologies/73679. Postmaster: Send all address changes to above address.

Titles in this Series

For a list of additional titles in this series, please visit:
http://www.igi-global.com/book-series/advances-environmental-engineering-green-technologies/73679

Convergence of Cloud Computing, AI, and Agricultural Science
Avinash Kumar Sharma (ABES Institute of Technology, India) Nitin Chanderwal (University of Cincinnati, USA) and Rijwan Khan (ABES Institute of Technology, India)
Engineering Science Reference • © 2023 • 375pp • H/C (ISBN: 9798369302002) • US $265.00

Global Industrial Impacts of Heavy Metal Pollution in Sub-Saharan Africa
Joan Nyika (University of Johannesburg, South Africa) and Megersa Olumana Dinka (University of Johannesburg, South Africa)
Engineering Science Reference • © 2023 • 406pp • H/C (ISBN: 9781668471166) • US $255.00

Nanopriming Approach to Sustainable Agriculture
Abhishek Singh (Faculty of Biology, Yerevan State University, Armenia) Vishnu D. Rajput (Academy of Biology and Biotechnology, Southern Federal University, Rostov-on-Don, Russia) Karen Ghazaryan (Faculty of Biology, Yerevan State University, Armenia) Santosh Kumar Gupta (National Institute of Plant Genome Research, India) and Tatiana Minkina (Academy of Biology and Biotechnology, Southern Federal University, Rostov-on-Don, Russia)
Engineering Science Reference • © 2023 • 453pp • H/C (ISBN: 9781668472323) • US $245.00

Handbook of Research on AI-Equipped IoT Applications in High-Tech Agriculture
Alex Khang (Global Research Institute of Technology and Engineering, USA)
Engineering Science Reference • © 2023 • 473pp • H/C (ISBN: 9781668492314) • US $335.00

Contemporary Developments in Agricultural Cyber-Physical Systems
G.S. Karthick (Department of Software Systems, PSG College of Arts and Science, India)
Engineering Science Reference • © 2023 • 293pp • H/C (ISBN: 9781668478790) • US $240.00

Perspectives on Ecological Degradation and Technological Progress
Veli Yilanci (Faculty of Political Sciences, Canakkale Onsekiz Mart University, Turkey)
Engineering Science Reference • © 2023 • 342pp • H/C (ISBN: 9781668467275) • US $225.00

IGI Global
PUBLISHER of TIMELY KNOWLEDGE

701 East Chocolate Avenue, Hershey, PA 17033, USA
Tel: 717-533-8845 x100 • Fax: 717-533-8661
E-Mail: cust@igi-global.com • www.igi-global.com

Table of Contents

Preface xiv

Chapter 1
Deep Learning Techniques for Smart Agriculture Applications 1
Ankita Mishra, Vellore Institute of Technology, India
Sourik Banerjee, Vellore Institute of Technology, India
Brijendra Singh, Vellore Institute of Technology, India

Chapter 2
Review Work of Automation Agricultural Robot System Using Machine Learning and Deep Learning 24
R. Felshiya Rajakumari, Department of Electrical and Electronics Engineering, Karpagam Institute of Technology, Coimbatore, India
M. Siva Ramkumar, Department of Electrical and Electronics Engineering, Karpagam Academy of Higher Education, Coimbatore, India

Chapter 3
Applications of Deep Learning and Machine Learning in Smart Agriculture: A Survey 34
Amrit pal Kaur, Manipal University Jaipur, India
Devershi Pallavi Bhatt, Manipal University Jaipur, India
Linesh Raja, Manipal University Jaipur, India

Chapter 4
Plant Disease Classification in Segmented Images Using Computer Vision 58
Rajashri Roy Choudhury, Brainware University, India
Piyal Roy, Brainware University, India
Shivnath Ghosh, Brainware University, India

Chapter 5
Enhancing Tomato Fruit Detection and Counting Through AI-Enabled Agricultural Innovations 93
S Gandhimathi Alias Usha, Velammal College of Engineering and Technology, Madurai, India

Chapter 6
Harnessing Environmental Intelligence to Enhance Crop Management by Leveraging Deep Learning Technique 106
S. Gandhimathi Alias Usha, Velammal College of Engineering and Technology, Madurai, India

Chapter 7
Crop Prediction for Smart Agriculture Using Ensemble of Classifiers 124
Khushal Kindra, Vellore Institute of Technology, Chennai, India
Bhuvaneswari Amma N. G., Vellore Institute of Technology, Chennai, India

Chapter 8
An Integrated Approach for Selection and Design of Sustainable Farmers' Protective-Hat in India 142
Suchismita Satapathy, KIIT University, India

Chapter 9
Smart Crop Protection System From Wild Animals Using Artificial Intelligence 173
Shailaja S. Mudengudi, Tontadrya College of Engineering, India
Muktha S. Patil, Tontadarya College of Engineering, India
Neetha S. Mudaraddi, Tontadarya College of Engineering, India

Chapter 10
GUI-Based End-to-End Deep Learning Model for Corn Leaf Disease Classification 192
G. Revathy, SASTRA University (Deemed), India
J. Jeyabharathi, Kalasalingam Academy of Research and Education, India
Madonna Arieth, Sreenivasa Institute of Technology and Management Studies, India
A. Ramalingam, SRI Manakula Vinayagar Engineering College (Autonomous), India

Chapter 11
A Transfer Learning Approach for Detecting Plant Leaf Diseases With Convolutional Neural Networks 214
P. Valarmathi, Vellore Institute of Technology, Chennai, India
N. G. Bhuvaneswari Amma, Vellore Institute of Technology, Chennai, India
Vasu Bhasin, Vellore Institute of Technology, Chennai, India

Compilation of References 225

About the Contributors 253

Index 256

Detailed Table of Contents

Preface..xiv

Chapter 1
Deep Learning Techniques for Smart Agriculture Applications 1
Ankita Mishra, Vellore Institute of Technology, India
Sourik Banerjee, Vellore Institute of Technology, India
Brijendra Singh, Vellore Institute of Technology, India

With an emphasis on the rapid and accurate diagnosis of plant and fruit diseases, researchers have been looking into sustainable agriculture utilizing cutting-edge deep learning techniques. The objective is to show how effective deep learning algorithms can revolutionize the agricultural industry. Automated illness detection is the main area of focus, where advances in image processing and computer vision techniques enable precise and quick identification while lowering labor requirements and associated costs. In order to identify plant and fruit diseases in a sustainable manner, this project intends to explore the possibilities of deep learning algorithms in detecting diseases from the leaves of agricultural plants using pre-trained deep convolutional neural network. This book chapter provides informative information on the use of deep learning in smart agriculture and a significant resource for researchers, professionals, and students interested in sustainable farming and intelligent agricultural systems.

Chapter 2
Review Work of Automation Agricultural Robot System Using Machine Learning and Deep Learning 24
R. Felshiya Rajakumari, Department of Electrical and Electronics Engineering, Karpagam Institute of Technology, Coimbatore, India
M. Siva Ramkumar, Department of Electrical and Electronics Engineering, Karpagam Academy of Higher Education, Coimbatore, India

An occupation based on automatic system using machine learning and deep learning techniques is developed in this chapter. The agricultural land and automatic systems are worked mutually to defeat the concern by integration with solar panel. This concern is resolved by the automatic system and it is employed very well for the farming procedures, and the automatic technology is cooperative with a high-quality vital role in engineering, medical field, military, etc. The major concern of the manuscript transaction the farming seeds is to lessen the chap authority interference. In this study, humidity sensor designates the earth irrigate contented in the farming field. In this chapter the complete works review with the automatic system by using machine learning and deep learning techniques for farming and weeding the seeds.

Chapter 3
Applications of Deep Learning and Machine Learning in Smart Agriculture: A Survey 34
Amrit pal Kaur, Manipal University Jaipur, India
Devershi Pallavi Bhatt, Manipal University Jaipur, India
Linesh Raja, Manipal University Jaipur, India

Machine learning (ML) and deep learning can be used in the smartest way possible to improve productivity in agriculture. The Food and Agriculture Organization's research shows that the crop's production is rising. One of the finest methods to monitor agricultural yield is through smart agriculture. Applications of ML and deep learning help to discover and resolve problems that crop growth encounters. In agriculture, the production of crops can be enhanced by applying machine learning and deep learning methodologies. These methods demonstrate the rapid advancement of artificial intelligence in the agriculture sector. The idea of "smart farming" keeps an eye on all processes, disease prediction, and agricultural pests. ML is used to extract meaningful information from huge datasets. Deep learning evaluates structural characteristics, meteorological data, and climatic factors to help anticipate agricultural diseases using practical and economical techniques. The deep-learning techniques enhance agricultural research's capacity to sense the overall classification of agriculture.

Chapter 4
Plant Disease Classification in Segmented Images Using Computer Vision.......58
Rajashri Roy Choudhury, Brainware University, India
Piyal Roy, Brainware University, India
Shivnath Ghosh, Brainware University, India

Agriculture productivity has a significant impact on the lives of people and economies because of the growing human population. In agriculture, plant diseases are a big problem since they result in severe crop losses and financial hardship for farmers. Traditional disease detection and categorization methods take a long time and are subjective, so automated and effective methods are required. Computer vision techniques have recently shown promise as tools for classifying plant diseases. To provide a precise and dependable system for disease detection and management, this article gives a thorough study on computer vision approaches for plant disease categorization. The research uses a variety of approaches, such as feature extraction, image pre-processing, and machine learning algorithms. Benchmark datasets are used for comparative study and performance evaluation of various methods. The outcomes show how effective computer vision techniques are at precisely diagnosing and categorising plant diseases.

Chapter 5
Enhancing Tomato Fruit Detection and Counting Through AI-Enabled Agricultural Innovations 93
S Gandhimathi Alias Usha, Velammal College of Engineering and Technology, Madurai, India

This chapter aims to enhance tomato fruit detection and counting in agricultural practices through AI-enabled innovations. Traditional manual methods for fruit detection and counting are labor-intensive and time-consuming. By exploiting AI technologies, such as computer vision and machine learning algorithms, this research proposes an automated system to accurately detect and count tomato fruits in real-time. The system utilizes image processing techniques and trained models to analyze images or videos captured in the field. The proposed approach has the potential to significantly improve efficiency, reduce costs, and increase accuracy in tomato fruit detection and counting, thereby benefiting the agricultural industry.

Chapter 6
Harnessing Environmental Intelligence to Enhance Crop Management by Leveraging Deep Learning Technique 106
S. Gandhimathi Alias Usha, Velammal College of Engineering and Technology, Madurai, India

This chapter aims to enhance crop management practices by harnessing environmental intelligence through the power of deep learning techniques. Efficient and sustainable crop management is crucial for meeting the increasing demand for agricultural products while minimizing environmental impact. In recent years, the integration of deep learning techniques with environmental data has shown great potential in improving crop management practices. The proposed approach involves training deep learning model Dense Net 121 to predict important crop management factors, including yield estimation, disease, and pest outbreaks. The models are trained using historical and real-time data, enabling them to adapt and respond to dynamic environmental conditions. By capturing complex patterns and interactions, deep learning models can provide valuable insights and recommendations to farmers, enabling them to optimize resource allocation, reduce input wastage, and improve overall crop productivity.

Chapter 7
Crop Prediction for Smart Agriculture Using Ensemble of Classifiers 124
Khushal Kindra, Vellore Institute of Technology, Chennai, India
Bhuvaneswari Amma N. G., Vellore Institute of Technology, Chennai, India

Nowadays due to the advancement in technology, smart agriculture is in the evolving stage. Agricultural farmers worldwide commonly utilize the process of cultivating and harvesting crops to produce food and fiber. Therefore, crop prediction is vital for smart agriculture and the proposed approach involves utilizing all the necessary resources to facilitate crop growth and maintenance. Crop cultivation used to be carried out based on farmers' actual experience. For farmers and agricultural decision-makers to make prompt and accurate judgments that will impact the caliber of agricultural harvests, advanced tools are required. A prediction has been supplanted by machine learning techniques as a result. To anticipate agricultural yield, the authors provide a prediction system based on an ensemble of machine learning classifiers. Also, they discovered that an ensemble technique provided more accurate prediction than using the currently available categorization algorithms.

Chapter 8
An Integrated Approach for Selection and Design of Sustainable Farmers' Protective-Hat in India 142
Suchismita Satapathy, KIIT University, India

Normally, the farmers in India are required to work in adverse climatic conditions while performing their agricultural activities. Though a number of preventive measures are available for the protection of farmers, the nominal and small scale farmers are still unable to bear these expenses. Therefore, an attempt was made in this study to the select the most appropriate material using Fuzzy-TOPSIS method to carry out the design of the head-cap for farmers in terms of parameters like stiffness, cost, durability, and availability in India. Further, for a sustainable design of these head-cap, the genetic algorithm (GA) as well as Sun Flower Optimization (SFO) method were use to obtain the best fitness values based on the anthropometric dimensions of farmers' head.

Chapter 9
Smart Crop Protection System From Wild Animals Using Artificial Intelligence 173
Shailaja S. Mudengudi, Tontadrya College of Engineering, India
Muktha S. Patil, Tontadarya College of Engineering, India
Neetha S. Mudaraddi, Tontadarya College of Engineering, India

The first major threat to the farmers is drought. Crop vandalization by animals is the second major threat after drought. Crops are vulnerable to animals. Therefore, it is very important to monitor the nearby presence of animals. The main aim of this project is to provide a better solution in order to resolve this problem. In this chapter, the authors proposed a method which could detect the presence of animal and offer a warning. They used a microcontroller and camera to detect the movement of animals, and send signal to the controller. It diverts the animal by producing sound; and, further, a signal is transmitted to GSM which gives an alert to the owner of the crop immediately. The proposed monitoring scheme is to provide an early warning about possible intrusion and damage by animals.

Chapter 10
GUI-Based End-to-End Deep Learning Model for Corn Leaf Disease Classification.......192
G. Revathy, SASTRA University (Deemed), India
J. Jeyabharathi, Kalasalingam Academy of Research and Education, India
Madonna Arieth, Sreenivasa Institute of Technology and Management Studies, India
A. Ramalingam, SRI Manakula Vinayagar Engineering College (Autonomous), India

Food security is a major problem worldwide. Ensuring that the crops produced are both safe and wholesome is crucial not only for people as the ultimate consumers of the crops, but also for farmers. Plant diseases are responsible for a significant percentage of crop losses. This alleviates the need for a fast and accurate model to discriminate and identify plants with diseases. The base chapter chosen aims to achieve the same through deep learning. The data set used in the work was obtained from Plant Village Dataset. The work customs deuce pre-trained models, EfficientNetB0 and DenseNet121, to citation the traits of the plants. The extracted traits are then fused together through concatenation to allow the model to read the more meaningful crop trait data. This also ensures that the different sets of feature data read by the two models compensate for any feature loss during extraction. It turns out that the above method gives better results associated to other models.

Chapter 11
A Transfer Learning Approach for Detecting Plant Leaf Diseases With Convolutional Neural Networks.......214
P. Valarmathi, Vellore Institute of Technology, Chennai, India
N. G. Bhuvaneswari Amma, Vellore Institute of Technology, Chennai, India
Vasu Bhasin, Vellore Institute of Technology, Chennai, India

Agriculture is an indispensable sector for the continuity of homo sapiens. In the Indian context where agriculture contributes 19.9 percent of GDP and engages almost 54.6 percent of the population, it requires great measures to be taken to avoid plant diseases. Plant diseases are difficult to track manually as that requires a lot of work and proficiency in plant diseases. It is mandatory to keep a check on at varying phases of crop development to recognize the disease at time. Digitalization as well as adopting technology is essential for agriculture and for the welfare of the farmer. This chapter concentrates on detecting plant diseases using convolutional neural network. The concept of transfer learning is used for classification. The proposed framework can identify various types of diseases efficiently.

Compilation of References 225

About the Contributors 253

Index 256

Preface

Welcome to *Machine Learning and Deep Learning for Smart Agriculture and Applications*. In this edited reference book, we delve into the fascinating realm of Artificial Intelligence (AI) as applied to Smart Agriculture. Our goal is to provide a comprehensive resource that explores the convergence of modern technology with agricultural practices, uncovering the transformative potential of Machine Learning (ML) and Deep Learning (DL) in this vital sector.

The landscape of agriculture is rapidly evolving, driven by the imperative of sustainable food production and the responsible management of resources. The advent of precision agriculture, digital farming, and innovative research concepts has paved the way for revolutionary approaches to cultivation. This book showcases the pivotal role played by geospatial technology, robotics, the Internet of Things (IoT), and AI-powered analytics in shaping the modern landscape of farming management.

Our journey takes us through an exploration of ML and DL's influence on various facets of agriculture. From automated irrigation systems to the use of agricultural drones for field analysis, from crop monitoring to animal health management, the applications are as diverse as they are promising. A particularly crucial area is the identification and tracking of plant diseases, where cutting-edge image processing and computer vision techniques are harnessed to enable rapid and accurate diagnoses.

As editors, we are excited to present this collection of insights, methodologies, and case studies contributed by experts at the forefront of AI in agriculture. This book is designed to serve as a guide for researchers, practitioners, and enthusiasts who wish to navigate the terrain of ML and DL in the context of agriculture. It provides a roadmap for leveraging advanced algorithms to enhance crop yields, optimize resource utilization, and foster sustainable practices.

The chapters within encompass a wide spectrum of topics, ranging from the analysis of agricultural data using ML methodologies, employing machine learning systems for crop and weed differentiation, bio-inspired algorithms, predictive modeling, disease classification through computer vision, and health monitoring of plants and fruits using ML and DL approaches. Additionally, we explore the realm of drone-based agriculture, pesticide impact detection, and the implementation of various deep learning architectures tailor-made for the demands of Smart Agriculture.

This book is a testament to the dynamic intersection of technology and agriculture, offering valuable insights and practical guidance. We extend our gratitude to the esteemed authors who have contributed their expertise and knowledge to this compilation. Their dedication to advancing the field is evident, and we hope that their contributions inspire new breakthroughs and innovative applications in the domain of Smart Agriculture.

We invite you, dear reader, to embark on this enlightening journey through the pages of "Machine Learning and Deep Learning for Smart Agriculture and Applications." May this book serve as a beacon of inspiration and knowledge, illuminating the path towards a more sustainable and productive future for agriculture.

CHAPTER OVERVIEW

Chapter 1: Deep Learning Techniques for Smart Agriculture Applications

In this chapter, the authors delve into the realm of sustainable agriculture, focusing on the rapid and precise diagnosis of plant and fruit diseases through advanced deep learning techniques. By exploring the potential of deep convolutional neural networks, the chapter highlights how these algorithms can revolutionize the agricultural industry. The primary emphasis is on automated disease detection, where cutting-edge image processing and computer vision techniques play a pivotal role in achieving accurate identification while minimizing labor and costs. This chapter offers invaluable insights for researchers, professionals, and students interested in the intersection of deep learning and smart agriculture, showcasing its transformative potential.

Chapter 2: Automation Agricultural Robot System Using Machine Learning and Deep Learning

Taking a step into the world of automation and technology-driven agriculture, this chapter focuses on the integration of Machine Learning and Deep Learning techniques for automated farming systems. The authors present an innovative approach to agricultural operations, utilizing solar panel integration and sensor-based technology. The chapter highlights the advantages of an automated farming system and its role in enhancing farming procedures, ranging from irrigation to seed planting. By reducing human intervention and leveraging advanced technologies, this chapter paves the way for more efficient and sustainable agricultural practices.

Chapter 3: Applications of Deep Learning and Machine Learning in Smart Agriculture: A Survey

Exploring the applications of Machine Learning (ML) and Deep Learning (DL) in the realm of smart agriculture, this chapter offers a comprehensive survey of their transformative potential. The authors emphasize the role of ML and DL in improving agricultural productivity and crop yield prediction. The chapter delves into the concept of "smart farming," encompassing disease prediction and pest control. By extracting meaningful insights from large datasets, ML and DL techniques are showcased as powerful tools to enhance agricultural practices. This survey serves as a valuable resource for understanding the wide-ranging applications of ML and DL in modern agriculture.

Chapter 4: Plant Disease Classification in Segmented Images using Computer Vision

Plant diseases pose a significant challenge to agricultural productivity, and this chapter focuses on innovative solutions through computer vision. The authors delve into the world of image processing, feature extraction, and machine learning algorithms to develop an accurate system for disease detection and categorization. By analyzing benchmark datasets and evaluating various techniques, the chapter underscores the potential of computer vision in precisely diagnosing and categorizing plant diseases. This chapter serves as a guide for researchers and practitioners seeking efficient methods for disease management in agriculture.

Chapter 5: Enhancing Tomato Fruit Detection and Counting Through AI-Enabled Agricultural Innovations

Traditional methods of tomato fruit detection and counting are labor-intensive and time-consuming. This chapter introduces AI-enabled innovations, harnessing computer vision and machine learning algorithms to automate and improve these processes. The authors present an automated system that analyzes images and videos captured in the field, accurately detecting and counting tomato fruits in real-time. By integrating technology into agricultural practices, this chapter offers a pathway to increased efficiency, reduced costs, and enhanced accuracy in tomato fruit detection, benefiting both farmers and the agricultural industry.

Chapter 6: Harnessing Environmental Intelligence to Enhance Crop Management by Leveraging Deep Learning Technique

Efficient crop management is essential for sustainable agriculture, and this chapter explores the fusion of deep learning techniques with environmental data. By training deep learning models to predict crucial crop management factors, such as yield estimation and disease outbreaks, the authors showcase the potential for technology-driven optimization. Historical and real-time data enable these models to adapt to dynamic conditions, providing farmers with valuable insights to improve resource allocation and overall crop productivity. This chapter highlights the transformative role of deep learning in enhancing agricultural practices.

Chapter 7: Crop Prediction for Smart Agriculture Using Ensemble of Classifiers

Addressing the evolving landscape of smart agriculture, this chapter focuses on the vital aspect of crop prediction. The authors propose an ensemble of machine learning classifiers to predict agricultural yields, highlighting the benefits of technology-driven decision-making. By harnessing machine learning techniques, the chapter underscores the importance of accurate predictions in agricultural planning and management. The chapter offers a pathway to empower farmers and decision-makers with valuable insights for optimizing crop production in the context of smart agriculture.

Chapter 8: An Integrated Approach for Selection and Design of Sustainable Farmers' Protective-Hat in India

In the context of adverse climatic conditions faced by farmers in India, this chapter delves into the design of a sustainable protective head-cap. Utilizing advanced methodologies such as Fuzzy-TOPSIS method and genetic algorithms, the authors propose an integrated approach for material selection and design. By considering parameters like stiffness, cost, durability, and availability, the chapter offers a solution to protect farmers while optimizing resource usage. This chapter showcases the intersection of technology, sustainability, and farmer well being in the context of agricultural innovation.

Chapter 9: Smart Crop Protection System From Wild Animals Using Artificial Intelligence

Wild animal interference poses a significant threat to crops, and this chapter presents an innovative solution through AI-enabled technology. The authors propose a system that detects animal presence and issues warnings to farmers. By combining microcontrollers, cameras, and GSM technology, the chapter showcases a comprehensive approach to mitigate crop damage. This smart protection system provides real-time alerts and demonstrates the potential of artificial intelligence in safeguarding agricultural yields.

Chapter 10: GUI Based End-to-End Deep Learning Model for Corn Leaf Disease Classification

Focusing on the critical issue of plant diseases, this chapter presents a GUI-based deep learning model for accurate disease classification in corn leaves. By leveraging deep learning architectures and image analysis techniques, the authors showcase an end-to-end solution for disease detection. The proposed model, incorporating pre-trained models and feature fusion, demonstrates promising results in identifying diseases. This chapter offers a comprehensive guide for researchers and practitioners seeking to enhance disease management and crop health in agriculture.

Chapter 11: A Transfer Learning Approach for Detecting Plant Leaf Diseases With Convolutional Neural Networks

Highlighting the importance of digitization and technology adoption in agriculture, this chapter presents a transfer learning approach for detecting plant leaf diseases. Leveraging convolutional neural networks, the authors emphasize the role of artificial intelligence in rapid and accurate disease identification. By utilizing historical data and applying transfer learning techniques, the chapter offers a solution to monitor and mitigate plant diseases effectively. This chapter provides valuable insights for implementing technology-driven disease detection strategies in modern agriculture.

IN SUMMARY

As we bring our exploration of "Machine Learning and Deep Learning for Smart Agriculture and Applications" to a close, we are reminded of the remarkable journey we have undertaken through the intersection of modern technology and age-old agricultural practices. The chapters within this book have illuminated the boundless possibilities that emerge when cutting-edge Machine Learning (ML) and Deep Learning (DL) techniques converge with the intricate world of agriculture.

From the earliest days of farming, humanity has sought innovative ways to enhance crop yields, mitigate risks, and foster sustainable practices. The digital era has ushered in a new wave of possibilities, empowering us to achieve these goals with unprecedented precision and efficiency. Our contributors, spanning diverse academic and professional backgrounds, have shared their expertise and insights to shed light on the transformative potential of AI in agriculture.

We have witnessed the power of deep learning algorithms revolutionizing disease detection, leading to accurate and timely diagnoses of plant and fruit diseases. The integration of automation and robotics powered by ML and DL has paved the way for efficient farming practices, optimizing resource allocation and minimizing human intervention. Moreover, our journey has led us through the realm of environmental intelligence, where deep learning models offer farmers valuable insights into crop management, pest outbreaks, and yield estimation.

The applications showcased in this book are not merely theoretical concepts; they represent a tangible shift towards a more sustainable and productive agricultural landscape. By harnessing the capabilities of AI, farmers and agricultural decision-makers are poised to make informed choices that impact both their livelihoods and the global food supply chain. The advancements in computer vision, machine learning, and deep learning have made it possible to tackle complex challenges, from disease classification to automated fruit detection, with unparalleled accuracy.

As editors, we extend our gratitude to the dedicated authors who have shared their knowledge and research, contributing to the compilation of this valuable resource. We hope that the insights presented within these pages inspire further innovation, collaboration, and exploration in the realm of smart agriculture.

In closing, "Machine Learning and Deep Learning for Smart Agriculture and Applications" stands as a testament to the potential of AI to reshape and redefine the future of farming. We invite readers, researchers, and practitioners to draw inspiration from these pages and embark on their own journey to harness the power of AI for a more sustainable and prosperous agricultural world.

Mohammad Farukh Hashmi
National Institute of Technology, Warangal, India

Avinash G. Keskar
Visvesvaraya National Institute of Technology, India

Chapter 1
Deep Learning Techniques for Smart Agriculture Applications

Ankita Mishra
https://orcid.org/0009-0004-7102-1455
Vellore Institute of Technology, India

Sourik Banerjee
https://orcid.org/0009-0002-3291-6999
Vellore Institute of Technology, India

Brijendra Singh
https://orcid.org/0000-0003-2608-3388
Vellore Institute of Technology, India

ABSTRACT

With an emphasis on the rapid and accurate diagnosis of plant and fruit diseases, researchers have been looking into sustainable agriculture utilizing cutting-edge deep learning techniques. The objective is to show how effective deep learning algorithms can revolutionize the agricultural industry. Automated illness detection is the main area of focus, where advances in image processing and computer vision techniques enable precise and quick identification while lowering labor requirements and associated costs. In order to identify plant and fruit diseases in a sustainable manner, this project intends to explore the possibilities of deep learning algorithms in detecting diseases from the leaves of agricultural plants using pre-trained deep convolutional neural network. This book chapter provides informative information on the use of deep learning in smart agriculture and a significant resource for researchers, professionals, and students interested in sustainable farming and intelligent agricultural systems.

DOI: 10.4018/978-1-6684-9975-7.ch001

1. INTRODUCTION

In order to overcome the difficulties associated with the identification of plant and fruit diseases within the context of sustainable agriculture, researchers have recently shown an increased interest in applying cutting-edge deep learning approaches. By increasing the efficiency and precision of disease detection, the agriculture business will undergo a transformation. Growing emphasis on utilizing cutting-edge technologies plays an important role in development of smart agriculture. The Internet of Things (IoT), big data, and deep learning frameworks are just a few of the cutting-edge technologies that academics are integrating to improve agricultural practices. We'll focus on how modern world technologies, like big data, machine learning, and deep learning frameworks, may boost agricultural systems' effectiveness and efficiency. As automation of disease detection is a crucial component of managing plant and fruit diseases, this will be our key area of concentration.

In order to identify plant and fruit diseases in a sustainable manner, this project intends to explore the possibilities of deep learning algorithms in detecting diseases from the leaves of agricultural plants since leaves are important in detecting plant diseases because they often show visible signs of sickness. Changes in leaf color, shape, and texture can indicate that a plant is infected. By closely examining the leaves, we can look for spots, discoloration, wilting, or unusual patterns that suggest the presence of a disease. Leaves also help us identify the specific disease and make decisions on how to manage it. Monitoring the progression of symptoms on the leaves helps us understand the effectiveness of treatments and interventions. Our study will concentrate on automating disease detection from the leaves using the Pre-Trained Deep Convolutional Neural Networks. We will also study these pre-trained models which are already trained with lots of images. We will emphasize the benefits and drawbacks of these technologies in the context of sustainable agriculture through the outcomes analysis.

Figure-1 depicts a deep learning framework for plant disease detection and control. By improving the effectiveness and precision of diagnosis of these leaves, the deployment of these cutting-edge technologies has the potential to transform agricultural practices. The approach used to examine the performance of deep learning algorithms in disease diagnosis through leaves will be covered in the following sections of this study. In order to emphasize the possible advantages and disadvantages of these technologies in the context of sustainable agriculture, we will present and analyze the results achieved.

The overall goal of this research project is to highlight how machine learning and deep learning algorithms can revolutionize the agricultural industry. By utilizing cutting-edge technologies, we can open the door for a more effective and long-

Figure 1. Deep learning framework for disease detection

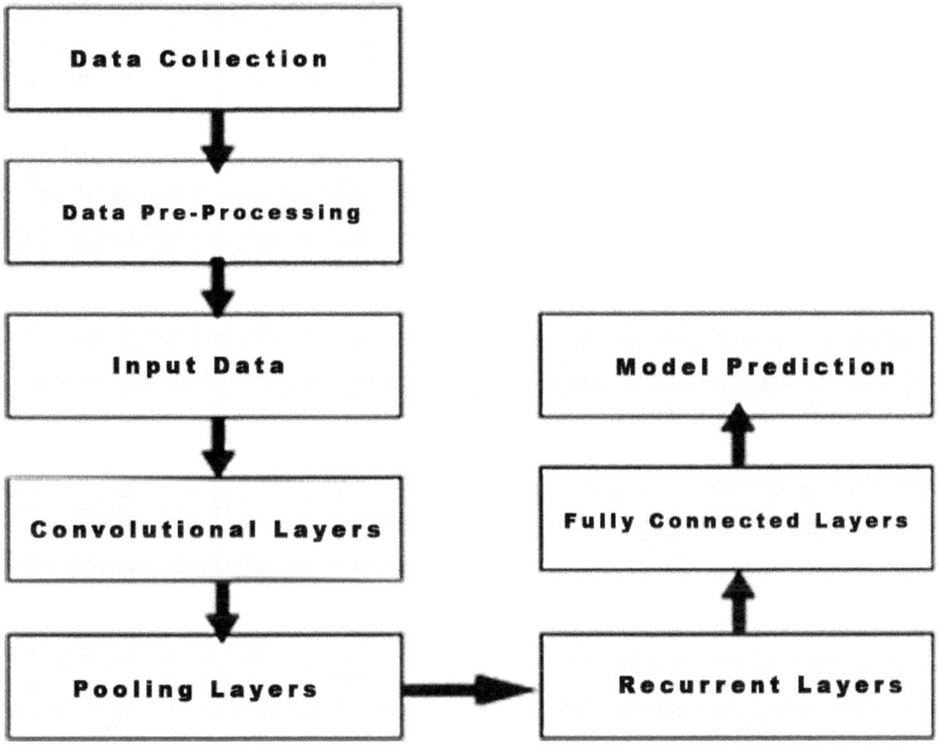

lasting method of managing plant and fruit diseases, which will benefit farmers, researchers, and the larger agricultural community.

Followed by introduction, rest of the book chapter is organized as follows. Related literature review presented in Section 2. Section 3 depicts methodology of the proposed work. Experimental setup and results are explained in Section 4. Finally, conclusion and future work is presented in Section 5.

2. REVIEW OF RELATED WORK

Three commonly used deep learning algorithms CNN, RNN, and GAN (Zhu et al., 2018) are introduced that are popular across a variety of fields. Although there are many sub-categories of deep learning algorithms, including VGGNet, ConvNets, LSTM, and DCGAN, the three most popular algorithms and their uses in agriculture are the main emphasis of this research. Additionally supplied are Python example codes for BP, CNN, RNN, and GAN, providing helpful implementation advice. Future potential in this field is examined, and the research on deep learning applications in agriculture is analyzed and summarized. The goal is to aid agricultural researchers in their understanding of deep learning algorithms and speed up their learning of the key deep learning approaches. This can thus make data analysis easier, advance related agricultural research, and effectively encourage the use of deep learning in this area. Although there are many commercial software frameworks available, a systematic overview of the main deep learning algorithms, including their principles, constraints, flowcharts, and sample programs, is lacking. This work attempts to fill that gap by offering agricultural researchers a thorough overview of deep learning techniques, allowing them to rapidly pick them up and efficiently apply them in their research.

Many aspects (Ünal, 2020) of smart farming are discussed, including the use of deep learning technologies in specific agricultural industries. The emphasis is on utilizing cutting-edge information technologies to increase productivity, monitor processes, and provide precise therapies. Deep learning offers improvements over earlier algorithms by enabling automatic feature extraction without human involvement, thanks to its capacity to extract latent structures from unlabeled and unstructured data. The articles appeared in 39 scientific publications after being retrieved from the Science Citation Index database. The report also provides data on the most prolific nations in terms of deep learning-based agricultural research as well as publication patterns. A major factor in the advancement of agricultural practices has been deep learning techniques. By extracting pertinent information from images, convolutional neural networks (CNNs) have demonstrated effectiveness in tasks like disease diagnosis, plant categorization, and weed identification. In order to estimate agricultural yield and identify pests, recurrent neural networks (RNNs) and their derivatives, such as LSTM, have been useful in capturing temporal connections in sequential data. These methods have allowed academics to fully utilize deep learning's promise for overcoming a variety of agricultural issues and improving farming practices.

The importance of agriculture (Ganatra & Patel, 2021) is emphasized in the opening section, which notes that about 65% of living things depend on agricultural products directly or indirectly. Due to issues including dwindling pollinators, crop diseases, inappropriate irrigation, technological limits, and water scarcity, the sector is challenged by population expansion and climate change. Deep learning, the Internet of Things, and computer vision must all work together to overcome these issues. As a field, precision agriculture integrates various technologies to address environmental issues, boost food yield, ensure sustainability, and raise standards. The most popular method for deep learning, particularly for computer vision problems, is explained as having a convolutional neural network (CNN) architecture. Convolutions, pooling layers, fully linked layers, activation functions, gates, memory cells, and other elements that depend on the network's structure are all included in the CNN architecture. Using CaffeNet for plant disease diagnosis as an example, the steps involved in feature extraction, dimensionality reduction, and classification/regression tasks within CNNs are explained.

The article (Maduranga & Abeysekera, 2020) outlines many agricultural applications for ML and IoT. Among these are animal tracks, soil management, disease management, crop and yield management, weed management, plant management, and crop management. ML algorithms may be used to follow animals on farms, diagnose and manage diseases, map and manage weeds, estimate crop yields, monitor and control environmental elements, and analyze soil qualities. The authors stress the significance of gathering trustworthy data from IoT devices in order to produce precise predictions and analyses when utilizing ML algorithms. They also talk about how ML and IoT technologies may work together, arguing that real-time data availability and connectivity offered by IoT devices can improve the efficacy of ML models. In addition to outlining novel concepts for fusing these technologies for improved precision agriculture, the study also underlines the possibilities of ML and IoT in smart agriculture. The authors examine current methods used in the subject and make recommendations for further study and use.

In order to meet the requirement for crop identification and detection, the authors (Zheng et al., 2019) present the CropDeep dataset, which focuses on vegetables and fruits directly related to precision agriculture. With at least 1500 samples for each of the 30 crop groups, the collection consists of more than 31,000 photos. Visual cameras placed in greenhouses, autonomous robots, and cellphones linked to IoT systems were used to capture the photos. CropDeep demonstrates domain-specific complexity by incorporating various crop components and growth stages as well as visually comparable sections of various species divided into several classes. The CropDeep dataset was used by the authors to test the effectiveness of various cutting-edge deep-learning frameworks for classification and detection. They set out to find architectures that could handle information quickly and accurately. The CropDeep dataset, which answers the demand for domain-specific crop vision datasets with thorough annotations for deep-learning-based classification and detection in precision agriculture, is introduced in the conclusion of the paper. The dataset includes a sizable number of photographs taken in greenhouses with various cameras, allowing for a more accurate benchmark for agricultural operations.

The report (Condran et al., 2022) emphasizes the inadequacy of current deep-learning networks for agricultural tasks such as crop management, despite their state-of-the-art performance in other academic domains. The main cause is the lack of publicly available benchmark datasets created especially for agricultural missions. To overcome this limitation and enable the creation of more potent deep-learning models for agriculture, adequate crop datasets must be built utilizing a variety of collection methods. The CropDeep dataset is distinctive in that it contains several sections and stages of a certain vegetable's or fruit's growth as well as visually comparable parts of various species. When compared to open datasets obtained from social media or the internet, precision agriculture is more sophisticated, as seen by the classification of various classes for leaves and flowers, for instance. On the CropDeep dataset, the authors ran tests utilizing a number of cutting-edge deep-learning frameworks to assess classification and detection performance. The key goal was to find architectures that were both quick and accurate enough to work with the sensors and machinery used in precision agriculture.

In particular, plant and crop classification using DL applications (Magomadov, 2019) in smart agriculture is the main topic of the research. DL models are employed in robotic harvesting, pest management, disaster monitoring, and yield prediction. Examples of DL models that have been used to plant disease detection, weed identification, and plant categorization are given in the study. Results reveal that CNN outperforms other machine learning algorithms for plant discrimination, and DL outperforms manual plant disease identification. In order to describe DL's capacity to process data in a nonlinear manner using hierarchical neural networks, the paper first gives an overview of the technology. Convolutional neural networks (CNNs) are emphasized as a crucial DL technique for tasks including face recognition, audio recognition, and natural language processing. The DL architecture, which consists of fully connected and convolutional layers of CNN, is shown. The report concludes by giving a summary of DL-based technologies and how they are used in agriculture. It emphasizes the utility of DL in numerous fields, such as fruit counting, disease detection, and plant classification.

By providing an in-depth analysis of current DL applications in agriculture (Santos et al., 2020), this literature assessment intends to close a gap while taking into account the unique hardware needs and real-time processing limitations imposed by complicated DL models. The findings offer important new perspectives on the development and potential of DL approaches in tackling agricultural problems and enhancing productivity, sustainability, and food security. The survey explores the idea of DL, highlighting designs that are frequently employed in agriculture, such as Convolutional Neural Networks (CNN), Fully Convolutional Networks (FCN), and Recurrent Neural Networks (RNN). While FCNs are used for semantic segmentation in agriculture and RNNs are best for temporal sequence tasks, CNNs have been used extensively for computer vision tasks in this sector. The assessment of DL applications in agriculture suggests that the most common areas of focus are disease identification, plant recognition, and land cover/weed identification.

In a fog computing-based smart agriculture setting, this study (Lee et al., 2020) suggests a technique termed Deep Learning Entrusted to Fog Nodes (DLEFN). The integration of DL with cloud and fog computing systems has been the subject of numerous studies. In order to show that high-performance mobile devices are capable of handling some of the DL processing, Lane et al. (Sa et al., 2018) introduced novel acceleration strategies for DL on mobile systems on a chip. In a cloud setting, Alsheikh et al. () coupled DL with the Apache Spark application, where mobile devices carried out DL operations and the cloud simultaneously learned from the data. Research is still needed to determine how to apply DL in IoT environments based on fog computing, despite these gains. Fog node devices frequently have different resources available in real-world circumstances because of things like incoming data size and network infrastructure. This problem is addressed by the proposed DLEFN algorithm, which determines the best DL layers for each fog node individually while taking into account their bandwidth and processing power. In comparison to the current assignment approaches, the proposed method supports more DL applications at fog nodes by lowering network traffic and computational overhead in the cloud.

The introduction to the part of the literature review discusses the importance of smart agriculture and how IoT technology has advanced it (Jin et al., 2020). The utilization of agricultural material data, early warning systems for pests and diseases, and monitoring plant production processes are just a few of the research that the authors mention in their discussion of agricultural IoT systems. In order to illustrate the whole process flow, the section also includes a schematic representation of the data collecting, communication, prediction, and application levels in an agricultural IoT system. The survey then concentrates on prediction techniques and how they might be used in time series analysis. Single approaches and combination methods, two groups of prediction methods, are covered. Physical analysis models and statistical models founded on concepts like autoregressive moving averages (ARMA), autoregressive integrated moving averages (ARIMA), and hidden Markov models (HMM) are examples of traditional approaches. These techniques are less suited for sophisticated agricultural applications since they frequently require prior information and presumptions. It demonstrates the value of deep learning techniques, the shortcomings of conventional approaches, and the necessity of reliable climate data forecasts in agriculture.

The literature study (Khan et al., 2021) demonstrates that the employment of UAVs in conjunction with machine learning and deep learning algorithms has considerably improved weed detection in agricultural regions. Support vector machines (SVM), random forest classifiers, convolutional neural networks (CNN), and object-based image analysis (OBIA) are just a few of the methods that have been investigated. These techniques have demonstrated their efficiency in various settings by achieving high accuracy in weed detection and mapping. Even with the advancements, it is still difficult to reliably identify weeds in crop fields, especially when crops and weeds have similar spectral properties and appearance. The resilience and effectiveness of weed detection technologies still need to be improved. In order to identify weeds and crops using RGB imagery recorded by UAVs, the study by Shahbaz Khan and colleagues improved the Faster R-CNN architecture based on deep learning. The outcomes show this method's potential to increase recognition accuracy and enhance decision-making in precision agriculture operations. The goal of ongoing research is to increase the accuracy and effectiveness of these systems in order to use fewer chemicals and support sustainable agriculture practices.

By integrating the knowledge, abilities, technology, and attitudes of the supply chain participants, sustainable agriculture supply chains (ASCs) play a crucial role in promoting sustainable practices (Sharma et al., 2020). Transferring knowledge is essential for motivating farmers to use sustainable agricultural methods. Due to reasons such as the depletion of water resources and fossil fuels, the scarcity of arable land, and consumer demand for transparent and sustainable food chains, ASCs are under increasing pressure to increase farming efficiency. Finally, in order to provide food security in the face of an increasing population, agriculture sustainability is crucial. ASCs are essential in advancing environmentally friendly agriculture methods. ASC efficiency, decision-making, and outcomes can all be improved with the integration of digital technology, especially AI and ML. The comprehensive literature evaluation carried out in this work helps to clarify how ML applications relate to sustainable ASC management and directs further study and real-world application efforts in this area.

The study (Sa et al., 2018) draws attention to the limits of earlier research on crop/weed semantic segmentation, which mostly focuses on processing single pictures for classification. UAV photos frequently have a small field of view and few spectral channels. It can be difficult to ensure global consistency and keep computation costs low when creating a map by segmenting photos and processing them one at a time. Furthermore, due to issues like constrained ground sample distances, reduced resolution from downsampling high-fidelity photos, and multispectral image alignment, producing a sizable and accurate vegetation map using a DNN is not simple. The authors suggest a new sliding window method that uses small tiles from multispectral ortho mosaic maps to overcome these issues. These tiles are calibrated radiometrically throughout the entire map and channel-wise aligned. Resolution loss is prevented by setting the tile size to correspond with the DNN input. For various classes (background, crop, and weed), the proposed model with 9 input channels outperforms the baseline model (SegNet with 3 channel RGB inputs). The literature review gives a summary of the current research in semantic segmentation using DNNs, UAV-based remote sensing for mapping vegetation, and the use of these methods in precision agriculture. The backdrop for the suggested framework is established, and the contributions and originality of the study are highlighted.

3. METHODOLOGY

In this section of the chapter, we have discussed some important topics like smart agriculture, deep learning, application in agriculture, and deep convolutional neural network (DCNN) models, namely DenseNet121 and MobileNetV2, for the prediction of diseases in agricultural plants based on leaf images.

3.1 Smart Agriculture

Precision farming, commonly referred to as smart farming, is a concept that incorporates the integration of cutting-edge technologies and data-driven methodologies to optimize farming practices and raise agricultural productivity. It aims to boost agriculture's productivity, sustainability, and profitability by utilizing cutting-edge technologies such as artificial intelligence (AI) and big data analytics. Utilizing a network of sensors and devices, smart agriculture gathers information on a variety of environmental factors, including temperature, humidity, soil moisture, and nutrient levels. Real-time monitoring and data collecting is made possible by these gadgets, giving decision-makers useful information.

3.2 Deep Learning

To forecast agricultural yields, deep learning models can analyze historical data, such as weather patterns, soil properties, and crop characteristics. Farmers that use this knowledge can use resources more efficiently, anticipate market demands, and make wise judgments. A branch of artificial intelligence (AI) known as "deep learning" focuses on teaching deep neural networks to perform complex tasks by mimicking how the human brain works. It has achieved tremendous advancements in several fields, including speech recognition, natural language processing, and computer vision, and it has the potential to fundamentally transform agriculture. Deep learning algorithms excel in discovering representations and patterns automatically from massive volumes of data, which enables them to make precise predictions and classifications.

3.3 Applications of Deep Learning in Smart Agriculture

Smart agriculture may greatly increase agricultural productivity and yield by maximizing resource use, putting focused interventions into place, and making data-driven decisions. Farmers may save money on water, fertilizer, and pesticides by utilizing smart agriculture to reduce input waste and optimize resource allocation. The qualities and attributes of plants that have been photographed by imaging techniques can be analyzed using deep learning models. These models can shed light on plant phenotypes by extracting variables associated with plant growth, leaf shape, or canopy development. Deep learning-based plant phenotyping provides the evaluation of crop health, growth stages, stress responses, and breeding choices. Utilizing deep learning's powers for agricultural production prediction, disease detection, weed management, and irrigation optimization would greatly advance smart agriculture. Deep neural networks and sophisticated data analytics can be used by farmers to make data-driven decisions, allocate resources more efficiently, and reduce risk. For deep learning to be widely used in smart agriculture, it is essential to address issues with data gathering, model interpretability, and scalability. Deep learning will continue to play a crucial role in revolutionizing conventional farming thanks to continued technological improvements and improved access to data.

3.4 Pre-Trained DCNN Model

Effective feature extractors are the pre-trained DCNN models. When used to identify plant species, spot illnesses, or assess crop health, the models learn to recognize different patterns, textures, and structures in photos. We can extract these learned features from the DCNN's intermediate layers and utilize them as input for jobs further down the line. Fine-tuning is an additional method used with pre-trained DCNN models in addition to feature extraction. The pre-trained model's weights are updated during fine-tuning using a smaller, domain-specific dataset. We can further modify the model to better comprehend and categorize agricultural patterns and

Figure 2. Deep learning in edge detection of a leaf
Bhattarai, S. (2018)

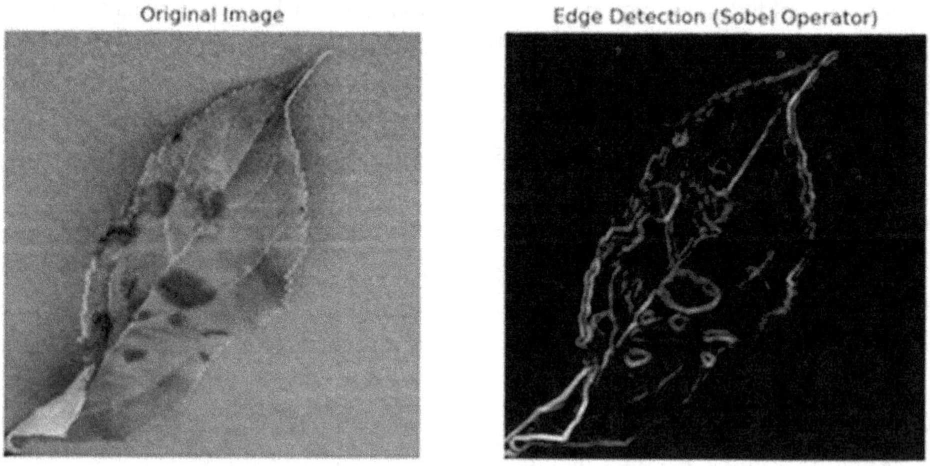

objects by training it on agricultural imagery or specific agricultural jobs. DCNN models that have already been trained include DenseNet, ResNet, InceptionNet, MobileNet, and many others. The quantity of the dataset, available computing power, and task difficulty are only a few examples of the variables that influence the model selection. In order to get correct results in Smart Agriculture applications, it is essential to choose the right model. Each model has its own architecture, depth, and performance characteristics. The pre-trained DCNN model can be deployed and incorporated into Smart Agriculture systems after being adjusted and validated. In order to do this, an effective pipeline must be created to preprocess input images, feed them into the model for inference, and then analyze the model's output to decide what to do or how to proceed.

3.5 MobileNet Model

For mobile and embedded devices with constrained processing resources, MobileNet is a compact CNN architecture. It maximizes accuracy while reducing the number of operations and parameters. MobileNet employs depthwise separable convolutions to factorise a regular convolution into a depthwise convolution and a pointwise convolution. This drastically lowers the computational cost while maintaining accuracy. The fundamental goal of MobileNet is to offer an architecture for neural networks that can achieve high accuracy while preserving a compact model and minimal computing cost. This qualifies it for deployment on hardware like smartphones, IoT devices, and embedded systems that have memory, computing power, and energy limitations.

Depth-wise separable convolutions are a technique that MobileNet uses to achieve efficiency. Depth-wise separable convolutions divide the convolution into two independent operations: depth-wise convolution and point-wise convolution. Traditional convolutional layers apply a full convolution operation to input channels.

1) **Depth-wise convolution:** This process convolves each input channel with a different filter to produce a series of output feature maps. By lowering the number of parameters used in the convolution procedure, this method drastically lowers the computing cost.
2) **Point-wise convolution:** The output feature maps from the depth-wise convolution are then subjected to a 1x1 convolution in this stage. By merging data from several sources, this technique improves the model's ability to recognize intricate patterns and correlations.

MobileNet strikes a fair balance between model size and accuracy by separating the spatial and channel-wise convolutions. In comparison to conventional convolutional neural networks like VGGNet or ResNet, it is substantially more efficient because it requires fewer parameters and calculations. To further manage the model size and computational cost, MobileNet also proposes a method dubbed "width multiplier". The target device's unique specifications can be taken into account when customizing the model using the width multiplier option, which scales the number of channels in each layer. The model size and computational cost can be greatly decreased while just slightly compromising accuracy by lowering the number of channels.

In terms of model size, computational complexity, and accuracy, MobileNet has been effective in striking a decent balance. Large-scale datasets like ImageNet and other picture classification projects have been shown to perform well in these situations. Its effectiveness and efficiency make it a popular choice for deploying deep learning models on devices with limited resources, enabling tasks like on-device machine learning applications, real-time object recognition, and image classification.

A sequence of depth-wise separable convolutions is used in MobileNet's design to create a neural network model that is both compact and effective. The architecture is explained in more detail below:

1) Input Layer: An image or tensor of shape (height, width, and channels) is used as the input for the MobileNet architecture.
2) Convolutional Layers: Traditional convolutional layers with modest filter sizes (3x3) and a stride of 2 make up the first layers of MobileNet. These layers are in charge of taking the input image's low-level features and extracting them.
3) Depth-wise Separable Convolutions: Deepness-wise separable convolutions are a common technique used by MobileNet to simplify computation. Two distinct operations make up a depth-wise separable convolution: both point-wise and depth-wise convolution.

Figure 3. Workflow of MobileNet

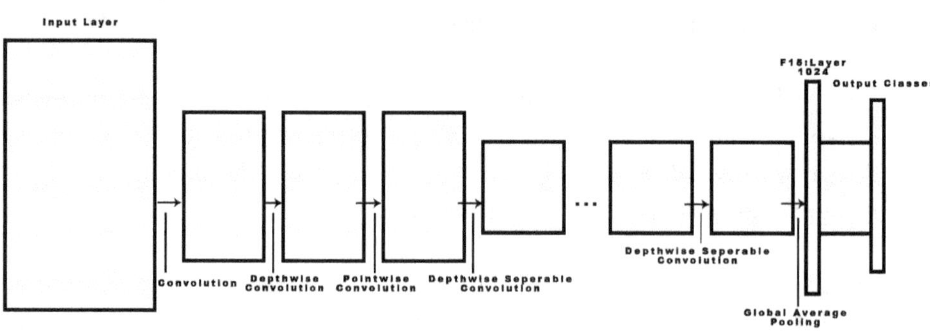

Depth-wise Convolution: A different filter is convolved with each input channel during this procedure. Each input channel is subjected to a single filter independently, producing a collection of feature maps with the same number of channels as the input. With fewer parameters, this process aids in the capture of geographical information.

Point-wise Convolution: To mix the data from several channels, a 1x1 point-wise convolution is used after the depth-wise convolution. The model's ability to recognize complicated patterns and relationships is improved by this process.

4) Bottleneck Layers: MobileNet also uses bottleneck layers, which first employ 1x1 convolutions to lower the number of input channels before implementing depth-wise separable convolutions. This keeps crucial information while reducing computational expense and model size.
5) Downsampling and Upsampling: Max pooling or stridden convolutions are two examples of downsampling layers that MobileNet uses to lower the spatial dimensions of the feature maps and expand the receptive field. In some MobileNet variations, the spatial dimensions may be increased by using upsampling layers like bilinear interpolation or transposed convolutions.
6) Fully Connected Layers and Softmax: Global average pooling, which determines the average value of each channel over the whole feature map, is often used in the final layers of MobileNet. The size of the feature map is decreased by this spatial pooling to a 1x1xC tensor, where C is the number of channels. The final class probabilities are then produced by passing the resulting tensor through fully connected layers and a softmax activation.

Through the use of hyperparameters like input resolution and width multiplier, MobileNet's design can be modified. The model can be made wider or thinner depending on the computational restrictions by using the width multiplier parameter, which scales the number of channels in each layer. The size of the input image depends on the input resolution, which can be changed to strike a compromise between precision and processing demands.

3.6 DenseNet Model

The dense connectivity pattern of DenseNet, also known as the Dense Convolutional Network, is well-known. Feature maps from earlier layers are integrated using summation or concatenation in conventional CNN systems. By immediately establishing feed-forward connections between every layer and every other layer, DenseNet adopts a different strategy. This extensive interconnectedness enhances model compactness, feature reuse, and gradient flow. Each layer in conventional convolutional neural networks is solely connected to the layer below it. However, in DenseNet, every layer is linked to every other layer in a feed-forward manner. Due to the direct and shorter paths that can be created by this dense connectivity, the vanishing gradient problem is lessened during back propagation. Additionally, the thick connections promote feature reuse across layers, which boosts accuracy and parameter efficiency.

In order to decrease the number of feature maps and regulate the spatial dimensions, DenseNet also uses transition layers between dense blocks. Batch normalization, a 1x1 convolutional layer, and down sampling techniques like average pooling are the main components of transition layers. These layers aid in information compression and offer a seamless transition between dense blocks. DenseNet's dense block is its most important part. A dense block is made up of several dense layers, each of which is connected to the ones above it. The feature maps from earlier layers are concatenated along the channel dimension within a dense block and used as inputs by later layers. Each layer has access to the feature maps of all lower layers thanks to this deep interconnectedness, enabling rich information sharing and feature reuse.

DenseNet's excellent parameter efficiency is a noteworthy advantage. In comparison to traditional architectures, the number of needed parameters is decreased because of the dense connectivity, which allows each layer to receive feature maps from all preceding levels. DenseNet is a good choice for applications with little training data or in resource-constrained settings because of its parameter efficiency. DenseNet has achieved state-of-the-art performance on a variety of computer vision tasks, including image classification, object identification, and image segmentation. It is an effective architecture for deep learning applications because of its dense connectivity and parameter effectiveness.

The architecture of DenseNet consists of dense blocks and transition layers, which are repeated throughout the network. Here is a detailed overview of the DenseNet architecture.

1) Input Layer: A tensor of shape (height, width, and channels) or a picture is used as the input to the DenseNet architecture.
2) Initial Convolutional Layer: The architecture begins with a single convolutional layer that has a stride of two and a modest filter size. This layer decreases the input's spatial dimensions and performs basic feature extraction.
3) Dense Blocks: The fundamental elements of DenseNet are dense blocks. Multiple thick layers make up each dense block. The exact DenseNet variation (such as DenseNet-121, DenseNet-169, or DenseNet-201) determines the number of dense layers within a dense block. Features are extracted and transformed at each dense layer. Dense Layers: Each dense layer in a dense block performs according to a predetermined pattern. Batch normalization comes first, then a ReLU activation function. The concatenated feature maps from all preceding

Figure 4. Workflow of DenseNet model

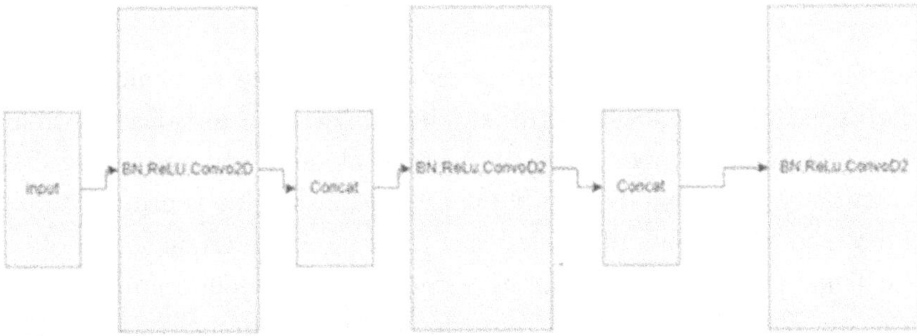

layers inside the same dense block are then subjected to a 3x3 convolutional layer. The subsequent dense layer is then fed with the output of this convolutional layer and the input feature maps from the dense layer.

4) Transition Layers: To limit the expansion of the number of feature maps and condense the spatial dimensions, transition layers are placed between dense blocks. Batch normalization, a 1x1 convolutional layer, and down sampling techniques like average pooling are the standard components of a transition layer. Down sampling techniques minimize the spatial dimensions while the 1x1 convolutional layer aids in reducing the number of feature mappings.
5) Global Average Pooling and Classification Layer: A global average pooling layer is used after the final dense block to combine the spatial data into a fixed-size feature vector. The final classification is carried out by feeding the pooled feature vector into a fully connected layer, followed by a softmax activation function.

A succession of dense blocks and transition layers built on top of one another, with an initial convolutional layer and a global average pooling layer at the ends, can be used to represent the overall architecture of DenseNet. Improved gradient flow, feature reuse, and parameter efficiency are just a few benefits offered by the DenseNet architecture, which help to explain why it performs so well in a variety of computer vision tasks.

3.7 Advantages of Pre-Trained DCNN Model

Here are some of the advantages of Pre-trained DCNN Models.

i. Transfer Learning: They have mastered the ability to recognize numerous patterns and representations and to extract relevant characteristics from images. Using a pre-trained model, you can use the model's knowledge and feature extraction abilities to a distinct but connected task. This is useful when you just have a little amount of labeled data to complete a certain activity.
ii. Save Training Time and Resources: Deep learning model training on strong hardware over a long period of time can be computationally expensive and time-consuming. As they have previously undergone considerable training on significant datasets, pre-trained models do not require additional training. You can save a lot of time and computing power by using a pre-trained model.
iii. Better Generalization: They can recognize common objects and concepts because they have learned a variety of features and patterns from their training data. Pretrained models can perform well on new and untried data because of their generalization capability, which makes them an excellent starting point for a variety of applications.
iv. Domain Adaptation: pre-trained models are better able to adapt to many domains and tasks as a result of exposure to diverse data. Even if there is little labeled data available for a particular task or domain, a pre-trained model can nevertheless offer useful knowledge and feature extraction skills.
v. Knowledge Transfer: We can apply this information to your particular activity by employing a pre-trained model, taking advantage of the model's comprehension of visual representations.
vi. Improved Performance: We can make use of a pre-trained model's high-performance abilities and possibly get greater outcomes than you would if you started from scratch or used simpler models.

3.8 Drawbacks of Pre-Trained DCNN Model

Here are some limitations of Pre-Trained DCNN Models.

i. Limited Flexibility: The architectures and learned representations of pre-trained models may not always completely match the requirements of your particular assignment because they were trained on specific tasks or domains. Due to their rigidity, they may be less adaptable and may need more adjusting or customizing to work at their best.

ii. Domain Mismatch: The performance of the pre-trained model could not generalize successfully if your assignment demands specialized knowledge or involves a vastly different subject. It may be more beneficial to fine-tune or start from scratch when using domain-specific data in some situations.
iii. Compatibility Issues: Different formats, such as TensorFlow, PyTorch, or Keras models, are frequently available for pre-trained models. If you're using a different deep learning framework or if the pre-trained model is not readily available in the format you need, this may lead to compatibility problems. It can be difficult and time-consuming to install compatibility layers or convert pre-trained models to the desired format.
iv. Model Size and Complexity: The file size and parameter count of pre-trained models might be rather enormous. If you have a shortage of storage space or processing power, this could be a problem. Furthermore, it could be more difficult to interpret or adapt particular components to meet your needs due to the model architecture's complexity.
v. Overfitting and Bias: Models that have already been trained may have biases towards the data they used to train them, which could affect predictions or representations. Furthermore, deploying a pre-trained model without sufficient fine-tuning or regularization methods might result in overfitting, where the model performs poorly on unobserved data or fails to generalize successfully.
vi. Dependency on Pretraining Data: The caliber and representativeness of the pretraining data have a significant impact on the performance of pre-trained models. The pre-trained model may inherit these restrictions and fail to perform at its best if the pretraining data is biased, noisy, or does not sufficiently represent the specific properties of the target task.

4. EXPERIMENT AND RESULTS

Due to its potential to completely alter agricultural practices, deep learning has drawn a lot of interest in the subject of smart agriculture. The theoretical framework used in this research to solve numerous issues in smart agriculture makes use of deep learning techniques. Our study will concentrate on automating disease detection from the leaves using the Pre-Trained Deep Convolutional Neural Networks. The dataset used in the experiment was obtained from Kaggle,() and the study's novel feature is the use of various deep-learning models. The suggested framework seeks to improve the effectiveness, precision, and sustainability of agricultural practices by analysis of the dataset and use of cutting-edge deep learning algorithms. Theoretical components of the proposed work, such as data collection and preprocessing, model selection, and prospective applications in smart agriculture, are discussed in this paper.

4.1 Data Collection and Pre-Processing

At first, we collected a dataset of 87K RGB images of healthy and diseased crop leaves from Kaggle database. For the dataset's compatibility with deep learning models, a standardized format is used to retrieve it. 38 separate classes are used to categorize these photos, which show various crop illnesses and health circumstances. The dataset is a useful tool for researching and creating remedies for crop health problems. Applying offline augmentation techniques improves the dataset's diversity and quality. These methods alter the source photographs in numerous ways, including rotation, flipping, resizing, and cropping. The dataset is expanded by creating more samples with enhanced variability, which improves model training and performance. The dataset is split into two primary subsets following the augmentation procedure: a training set and a validation set. 80% of the enhanced photographs are in the training set, while the remaining 20% are in the validation set. Through this divide, it is made sure that the models are trained on a sizeable section of the dataset and tested on a different portion, allowing for a trustworthy evaluation of their performance. The creation of a test set with 33 additional photos is also done in a separate directory. These pictures are not part of the supplemented dataset or the original dataset. This test set's objective is to assess how well the trained models generalize to new data. We can evaluate the algorithms' efficacy in correctly forecasting the health status of crop leaves using this different set of photos.

4.2 Preparing the Pre-Trained DCNN Model

In this study, we compare the efficacy of two pre-trained deep convolutional neural networks (DCNN) models for fruit and plant disease detection: DenseNet121 and MobileNet. By contrasting the accuracy levels attained by various models, we may determine which model produces the greatest outcomes for the purposes of our

Figure 5. Number of images in 38 classes in the dataset

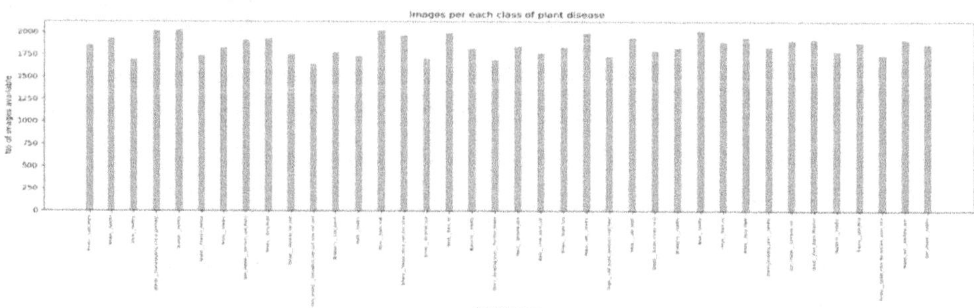

Figure 6. Sample data from each class
Bhattarai, S. (2018)

particular application. Modern DCNN models like DenseNet121 and MobileNet have all shown good performance in a variety of computer vision applications, including picture classification. These models have already been pre-trained on big image datasets, like ImageNet, and have mastered the art of extracting incredibly useful information from pictures. Transfer learning will be used to get the pre-trained models ready for our illness detection assignment. We can use transfer learning to apply the information that these models have amassed on massive datasets to our particular domain. We can modify the previously trained models to recognize disease patterns unique to our application by fine-tuning them using data on plant and fruit diseases. With the appropriate weights acquired from pre-training on ImageNet, we will initialize the pre-trained models. To match the number of disease classes in our dataset, the fully connected layers at the top of the network will be replaced or retrained. The models can acquire the discriminative features necessary for our illness detection task through this method. To fine-tune the models on our dataset during the training phase, we will use the proper optimization techniques, such as stochastic gradient descent (SGD) or Adam. The validation set will be utilized for model evaluation and hyperparameter adjustment once the models have been trained using the training set.

4.3 Results

In this experiment, we trained and evaluated two pre-trained deep convolutional neural networks (DCNN) models, DenseNet121 and MobileNetV2, on a large dataset consisting of agricultural plant leaf images to predict diseases. The goal is to compare the performance of these models in terms of accuracy and loss.

After training the models, we obtained promising results. DenseNet121 achieved an accuracy of 92.49% with a loss of 22.69%, while MobileNetV2 achieved a slightly higher accuracy of 93.05% with a loss of 20.81%. These results indicate that both models performed well in accurately predicting diseases in agricultural plants, with MobileNetV2 demonstrating a slightly better performance. The high accuracies obtained by both models indicate their ability to effectively learn and classify the different disease patterns present in the leaf images. This suggests that DCNN models are highly capable of extracting relevant features from agricultural plant leaf images and utilizing them for accurate disease prediction.

The comparison of accuracy and loss between DenseNet121 and MobileNetV2 reveals that MobileNetV2 outperformed DenseNet121 in terms of accuracy, albeit with a slightly lower loss. This implies that MobileNetV2 achieved a higher proportion of correct predictions while also minimizing the overall error.

The results of this experiment highlight the potential of utilizing pre-trained DCNN models, such as DenseNet121 and MobileNetV2, for disease prediction in agricultural plants. The obtained accuracies demonstrate the feasibility of employing such models in practical applications, where early disease detection can greatly assist in mitigating crop losses and improving agricultural productivity.

It should be noted that these results are specific to the dataset and experimental setup used in this study. Further investigations and evaluations on diverse datasets and real-world scenarios are recommended to validate the generalizability and robustness of these models for agricultural plant disease prediction.

5. CONCLUSION

In conclusion, deep learning techniques have emerged as powerful tools for smart agriculture applications, revolutionizing the way we address various challenges in the agricultural sector. The use of deep learning models, such as DenseNet121 and MobileNetV2, has showcased their potential in disease prediction and management in agricultural plants based on leaf images. These models have demonstrated high

Table 1. Model performance comparison

Model	Accuracy (%)	Loss (%)
DenseNet121	92.49	22.69
MobileNetV2	93.05	20.81

Figure 7. Accuracy and loss comparison of the two models

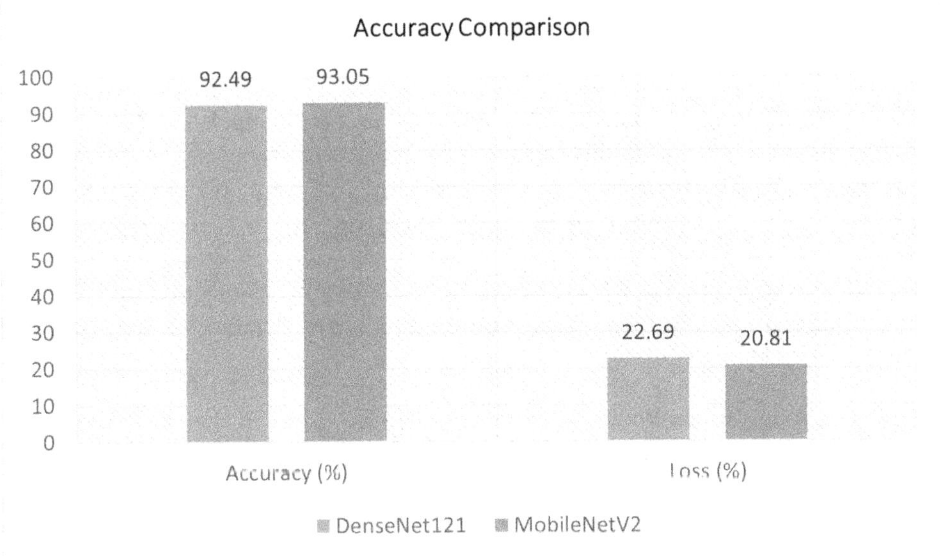

accuracies, indicating their ability to effectively learn and classify complex patterns associated with plant diseases. By enabling early disease detection, deep learning techniques have the potential to significantly reduce crop losses and improve agricultural productivity.

Moreover, deep learning techniques offer several advantages for smart agriculture applications. They can handle large-scale datasets, learn intricate relationships from raw data, and adapt to diverse agricultural environments. With their ability to extract meaningful features automatically, deep learning models alleviate the burden of manual feature engineering, making them versatile and efficient for different plant disease recognition tasks.

Our experiment specifically evaluated the performance of DenseNet121 and MobileNetV2 models in disease prediction based on leaf images. Both models exhibited strong predictive capabilities, with MobileNetV2 demonstrating a slightly higher accuracy compared to DenseNet121. This experiment further solidifies the potential of deep learning techniques for accurate disease diagnosis and highlights the importance of selecting appropriate models for specific agricultural applications.

While deep learning techniques hold great promise for smart agriculture, there are still challenges to address. Issues such as the limited availability of labeled datasets, interpretability of models, and model robustness in real-world conditions require continued research and development efforts.

REFERENCES

Bhattarai, S. (2018). *New Plant Disease Dataset.* Kaggle. https://www.kaggle.com/datasets/vipoooool/new-plant-diseases-dataset

Condran, S., Bewong, M., Islam, M. Z., Maphosa, L., & Zheng, L. (2022). Machine Learning in Precision Agriculture: A Survey on Trends, Applications and Evaluations Over Two Decades. *IEEE Access : Practical Innovations, Open Solutions, 10*, 73786–73803. doi:10.1109/ACCESS.2022.3188649

Ganatra, N., & Patel, A. (2021). Deep learning methods and applications for precision agriculture. Machine Learning for Predictive Analysis. *Proceedings of ICTIS, 2020*, 515–527.

Jin, X. B., Yang, N. X., Wang, X. Y., Bai, Y. T., Su, T. L., & Kong, J. L. (2020). Hybrid deep learning predictor for smart agriculture sensing based on empirical mode decomposition and gated recurrent unit group model. *Sensors (Basel), 20*(5), 1334. doi:10.339020051334 PMID:32121411

Khan, S., Tufail, M., Khan, M. T., Khan, Z. A., & Anwar, S. (2021). Deep learning-based identification system of weeds and crops in strawberry and pea fields for a precision agriculture sprayer. *Precision Agriculture, 22*(6), 1711–1727. doi:10.100711119-021-09808-9

Lee, K., Silva, B. N., & Han, K. (2020). Deep learning entrusted to fog nodes (DLEFN) based smart agriculture. *Applied Sciences (Basel, Switzerland), 10*(4), 1544. doi:10.3390/app10041544

Maduranga, M. W. P., & Abeysekera, R. (2020). Machine learning applications in IoT based agriculture and smart farming: A review. *Int. J. Eng. Appl. Sci. Technol, 4*(12), 24–27. doi:10.33564/IJEAST.2020.v04i12.004

Magomadov, V. S. (2019, December). Deep learning and its role in smart agriculture. *Journal of Physics: Conference Series, 1399*(4), 044109. doi:10.1088/1742-6596/1399/4/044109

Sa, I., Popović, M., Khanna, R., Chen, Z., Lottes, P., Liebisch, F., Nieto, J., Stachniss, C., Walter, A., & Siegwart, R. (2018). WeedMap: A large-scale semantic weed mapping framework using aerial multispectral imaging and deep neural network for precision farming. *Remote Sensing (Basel), 10*(9), 1423. doi:10.3390/rs10091423

Santos, L., Santos, F. N., Oliveira, P. M., & Shinde, P. (2020). Deep learning applications in agriculture: A short review. *Robot 2019: Fourth Iberian Robotics Conference Advanced Robotics, 1*, 139–151.

Sharma, R., Kamble, S. S., Gunasekaran, A., Kumar, V., & Kumar, A. (2020). A systematic literature review on machine learning applications for sustainable agriculture supply chain performance. *Computers & Operations Research, 119*, 104926. doi:10.1016/j.cor.2020.104926

Ünal, Z. (2020). Smart farming becomes even smarter with deep learning—A bibliographical analysis. *IEEE Access : Practical Innovations, Open Solutions*, *8*, 105587–105609. doi:10.1109/ACCESS.2020.3000175

Zheng, Y. Y., Kong, J. L., Jin, X. B., Wang, X. Y., Su, T. L., & Zuo, M. (2019). CropDeep: The crop vision dataset for deep-learning-based classification and detection in precision agriculture. *Sensors (Basel)*, *19*(5), 1058. doi:10.339019051058 PMID:30832283

Zhu, N., Liu, X., Liu, Z., Hu, K., Wang, Y., Tan, J., Huang, M., Zhu, Q., Ji, X., Jiang, Y., & Guo, Y. (2018). Deep learning for smart agriculture: Concepts, tools, applications, and opportunities. *International Journal of Agricultural and Biological Engineering*, *11*(4), 32–44. doi:10.25165/j.ijabe.20181104.4475

Chapter 2
Review Work of Automation Agricultural Robot System Using Machine Learning and Deep Learning

R. Felshiya Rajakumari
Department of Electrical and Electronics Engineering, Karpagam Institute of Technology, Coimbatore, India

M. Siva Ramkumar
Department of Electrical and Electronics Engineering, Karpagam Academy of Higher Education, Coimbatore, India

ABSTRACT

An occupation based on automatic system using machine learning and deep learning techniques is developed in this chapter. The agricultural land and automatic systems are worked mutually to defeat the concern by integration with solar panel. This concern is resolved by the automatic system and it is employed very well for the farming procedures, and the automatic technology is cooperative with a high-quality vital role in engineering, medical field, military, etc. The major concern of the manuscript transaction the farming seeds is to lessen the chap authority interference. In this study, humidity sensor designates the earth irrigate contented in the farming field. In this chapter the complete works review with the automatic system by using machine learning and deep learning techniques for farming and weeding the seeds.

1. INTRODUCTION

Cultivation is a significant rising division all through the planet area due to the mounting population. The goal of the labor in the gardening is to confront the output and the

DOI: 10.4018/978-1-6684-9975-7.ch002

superiority of agricultural to fetch it more victory in the farming to gather the increasing demand for the food [1]. Now due to population in the world, the planter capitalizes the production to cut the expenditure of seed, farming, etc., in the agriculture area [1] [2]. Farming automaton is one of the main key in machinery in this technology to perform the service resourcefully. Agricultural is the predictable technique and it is used at the present time knowledge for cultivation process [2]. In this manuscript the labor automates the employment work to uphold the humidity content soil.

Agriculture has changed throughout history as a result of modern machinery and technology. Utilizing these tools and technologies raises the standard of farming, decreases labor requirements and meets the demand for qualified agricultural workers. The roll-out of new equipment propels agriculture towards a new stage of growing crops.

If autonomous robots are developed, it may change the financial aspects of machine construction and lead to the development of smaller, slower machines. The swarm notion is in fact based on this removal of the driver overhead per car. The current generation of massive, powerful vehicles obviously outperforms fleets of slow, little robots in terms of productivity. The productivity gap, nevertheless, can only get smaller because there is still a lot of space for growth in the latter, even in the absence of a discovery or novel idea.

The way farming is done will undergo a profound and comprehensive revolution as a result of the advancements in agricultural robots, machine vision, and AI. By using comparison with the entire global agriculture industry, the fleet sizes and the total surface covered by new robots are still vanishingly small. The ground is slowly but steadily moving, so this shouldn't give the players a false sense of security. Artificial intelligence and robotics are enabling a revolution in cheap ultra precision that will eventually challenge long standing conventions in agricultural practices, the supply of agrochemicals and the design of agricultural machinery.

By using numerous data, machine learning and deep learning methods are applied to the automation agricultural robot. A huge enhance in food manufacture must be recognized as the global residents increases. It is complex for modern tools to assurance the reliable supply and features of food across the global without destructive normal environment [4] [7]. In agriculture, the use of machine learning has a wide range of possible uses. Farmers and agricultural experts are researching ways to increase crop yields, use less water, and detect pests and diseases by exploiting automated learning loans. Modern growers may benefit from data mining by using resources more effectively and producing food in a sustainable manner.

Due to the accessibility of intelligent systems, modern machines are becoming increasingly sophisticated and operating in a healthier and better manner. Important characteristics of contemporary AI include intelligent, advanced human speech, built-in strategic systems, intelligent routing, and self-driving cars. The majority of farming units are dealing with a skilled labour shortage and AI can have an important impact in encouraging the development of qualified workers that leads to higher harvesting and farm outputs. The rapid development of cognitive computing, AI and machine learning is transforming manual, analogue agriculture equipment and services into automated, digital ones. Due to the significance of its economy, agriculture plays an essential part in the majority of countries; as a result, it is regarded as crucial to development. The primary tasks of AI are believed to involve cognitive thinking, machine learning (ML), natural language processing, computer vision, etc. This work affords an about deep learning and machine learning techniques based solutions for agriculture field. The significance of both the applications created for the agricultural field and provide offer in determining the agricultural issues. The function of deep learning and machine learning performance in agriculture is still in its active phase. As a practice, we have created the offerings on ML and DL to solve the problems in the agricultural area.

Segment 1 contract with the Introduction work. Portion 2 presents the labor incentive of farming automatic system. Section 3 discuss with the employment of automatic system. Sector 4 states the Working of Autonomous Farming Seeding Robot. Segment 5 currents the automation of agricultural system by ML and DL. Segment 6 predicts the experimental analysis of prototype model work. Finally, sector 7 discusses the conclusion work. Portion 8 states the future scope work for automation agricultural system.

2. FARMING AUTOMATIC SYSTEM

In this process, two sensor techniques are used:

- Humidity sensor
- Ultrasonic sensor

The work is employed with the GSM module expertise, so the resulting work will proficiently act as an efficient manner (Xie & Liu, 2021). By means of this future expertise, it takes delivery in sequence to the cultivator to take the essential action to shun the tricky condition while cultivating the agricultural land.

3. EMPLOYMENT OF AUTOMATIC SYSTEM

- Production is little and occasion expenditure is gigantic.
- The deep learning and machine learning techniques are used to automate the agricultural robot.
- Farming area need to check regularly, the process will be monitor in the computer.

4. WORKING OF AUTONOMOUS FARMING SEEDING ROBOT

Agriculture farming is performing of modern art technology such as sensors, robotics for changing monotonous operation to computerized process. Latest achievements in agricultural robotics are used for sovereign tidy control, field reconnaissance and harvesting. It implies in research focus that agricultural robots edifice a small scale of work and drones to optimize the farming inputs to the land. Replacing man made power work with automations a growing trend in the field of agriculture. Robotics and Autonomous system is combining together to transform the global industries. By using these technologies in the world, it has the greatest impact in the agriculture field; hence it reduces the productivity farm. The advent of using autonomous farm in the agriculture field it reducers the input by using robot and it works as a more efficient way. The recent focus is based on agri-robotics it is used to identify the agriculture applications where the automation is been used as more efficient than a traditional human being [4]. By using this agri-robotics, it improves the life quality of farm workers by reducing the human work and tedious process in agriculture. Most of the current and recent agriculture technology falls into methods of smart farm such as: autonomous robots, drones and sensors. The aim purpose of moisture sensor work is, it will be senses the moisture level of the soil in the agriculture land. This sensor implements the work in the land and it produce the voltage reading which is directly proportional to the dielectric permittivity. In this field of agriculture, the sowing or seeding mechanism with long toothed gears actuated with motors [9] [10]. The structure of the seeding mechanism is partially divided into two portions and it act completely as a container for seeds. Successor holds electronics for automating and actuation and there is a gear at the back of the robot rotate. Ultrasonic is used for distance calculating the obstacles to indicate the robot to turn the direction for seeding the cultivation land Most commonly used-to detect hidden tracks, discontinuities in metals a water level detection [7]. The module of the ultrasonic sensor includes transmitter, receiver and control circuit. Ultrasonic receiver would stop timing when it receives the direction the reflected wave and it also senses the obstacles in the farming area.

Industrial systems are hampered by the lack of field security facilities and call for large-scale, long-term solutions, unlike robots in industries where human labour zones can be blocked when a machine is in use. Finding themes in records that do not follow customary or expected behaviour is referred to as anomaly detection. A detector of anomalies that makes use of the farming industry's uniform personality features is part of the backdrop and deep learning combo that defines the top performance configuration known as underground abnormal.

5. AUTOMATION OF AGRICULTURAL SYSTEM BY ML AND DL

The intersection of AI and automated learning is a field that is as widespread as energy in the present-day world, therefore its usage in robotics is likewise becoming increasingly important. Robots are being educated, and their accuracy is being grew, using precise automated learning techniques. Robots may acquire knowledge through neural networks how to carry out tasks like holding items, understanding relationships between spaces, recognizing objects, movement control, and other activities which enable them to communicate with the environment.

Cultivation is the stage of contemporary work such as sensors, automatic system for altering droning procedure to preset procedure by using machine learning and deep learning. Agricultural applications are exceptionally basic and proficient recognition to machine learning. When person knowledge is inadequate to knob complex responsibilities, machine learning techniques are classically organized. To estimate soil content like moisture eradication, crop production, machine learning techniques are used and also farming and weeding process can be utilized in agricultural field. A critical characteristic of the deep learning method, which develops the qualities of a farming field to the particular area for doing agriculture work, is its capacity to recognize unfamiliar equipments slightly than just a gathering of modern items [4] [7].

The current wave of technological advancements that combine robots and machine learning appear to be the most potent pairing yet. Every institution of human civilization is about to be upended by an entirely new era of automation. Robots powered by AI are thought to be more productive than those without the technology. For instance, the industrial sector is the greatest user of robots and further automation, which ensures validity, correctness, and minimal errors while saving time and human labor. One cannot put the safety of the location at risk with simple, traditional robots in such a crowded workspace. This is where machine learning and artificial intelligence are relevant. AI equips robots with sufficient vision technology and control of motion to more effectively. Similar to this, machine learning trains the robots so that they quickly evolve and learn from their own mistakes, negating the need for continuous human intervention and parallel effort. This guarantees robotics' flexibility. Along with these consequences, AI and ML undoubtedly increase the efficiency of production processes, especially for large labor-intensive businesses; they also expand the potential of robots.

The ability of the robots to perceive, recognize, localize, and conduct site-specific, intelligent action on individual plants is made possible by machine vision technology, which is a fundamental capability. Machine vision increasingly makes use of deep learning algorithms, which are frequently trained on image datasets with expert annotations. This allows the technology to perform significantly better than traditional algorithms, and occasionally even professional agronomists. Importantly, this strategy makes it possible to create a long-term technological roadmap that can be expanded to identify all varieties of crops and study the conditions that are related to them, such as water stress, disease, etc.

By means of this knowledge in the globe, it has the maximum collision in the cultivation field; hence it lessens the competence work. The arrival of work over agriculture area in the farming field it diminish the contribution by using robot and it works as a more efficient way. The fresh meeting point is based on agro-robotics it is used to spot the farming request where the mechanization is used as more capable than a conventional person creature [5] [10] [11]. By means of agro-robotics, it progress the existence the excellence of ranch recruits by dipping the creature employment and boring course in cultivation. The majority of the recent work and farming knowledge waterfall into technique of elegant ranch such as: automatic system.

Machine vision technology today has a good success rate for localizing and identifying various apparent fruit versus complex backdrops. Performance has increased as a result of the development of deep learning-powered image recognition technology. Additionally, there is a clear path for developing algorithms for novel fruit-environment combinations, allowing machine recognition and localization to be applied to a wide range of fruits. Other difficulties include selecting approach, robotic path scheduling, and robotic arm action control. There are algorithmic advancements in this area as well. More significantly, businesses are creating innovative end-effectors that can quicken delicate fresh fruit harvesting while reducing the computational burden.

DL-based agriculture robots result in increased productivity, healthier soil, and higher yield. An automaton employed in food production is called an agriculture robot. The key areas of industry where robots are now intervening are in the cultivation stage. Herbicides, crop digging up, tilling, ground evaluation, and remote sensing are typically used in early automaton or drone implementations in farming. Mechanization was built into the design of factory work robots such as forklifts, canisters without operators, and flock sheering robots. Vertical agriculture and industrial farming both try to grow crops in controlled circumstances that can either enhance plant quality or replicate a specific region's environmental conditions.

This enlargement has been possible due to mounting development of moisture and ultrasonic sensors and the reduction equipment costs due to mass production and innovative technologies. As a result, it shows that robotics has been successfully implemented in agricultural tasks to reduce the work load of the man power and optimize the process time.

6. EXPERIMENTAL ANALYSIS

A prototype arrangement intended for the farming robot is residential and experienced in the lab work. The course gears are used for prototype completion are scheduled in Table 1. Figure 1 designate the prototype model of automatic system. Figure 2 demonstrate the Seeding Mechanism for Agricultural Robot. Finally the model is equipped with agricultural robot for farming and seeding mechanism. The supply gives supremacy to the entire circuit, the agricultural robot gives best performance for farming and seeding mechanism.

This enlargement has been possible due to mounting development of moisture and ultrasonic sensors and the reduction equipment costs due to mass production and innovative technologies. As a result, it shows that robotics has been successfully implemented in agricultural tasks to reduce the work load of the man power and optimize the process time

7. CONCLUSION

In this work, Arduino based autonomous farming seeding robot for agricultural purpose are analyzed in detail. Automatic Cultivation is possible without human intervention. Seeding, Manuaring and supplying water is possible due to high production rate. Due to the presence of Ultrasonic sensors the robot moves throughout the field. This automatic way of sowing seeds using a robot reduces the labor requirement. Here the wastage of seeds is also been reduced to a greater extent. Here with the help of a robot the seeds are been dispensed in the soil in a proper sequence hereby reducing the wastage of seeds. This robot will help the farmers to do the farming process efficiently. In future, this work will carry out with the present condition of the neural networks, machine learning, and robotics in almost every sector, especially

Table 1. Prototype Parameters for Automatic Robot System

Apparatus	Parameters
LCD	16X2
Oscillator	4MhZ
Transformer	230V/12V
Regulator	LM7805
LM7812	
OP-AMP	LM358
Humidity Sensor	3.3-5V

Figure 1. Hardware prototype of Automatic System

Figure 2. Scattering Mechanism for Agricultural Robot

those in agriculture, healthcare, education, transportation, and a host of other fields. This implies that regardless of the most isolated regions, more powerful innovations and adequate awareness are constantly needed. Robotics technologies using AI and ML are expanding rapidly. In the coming generations, computational intelligence is going to become an imperative rather than an indulgence. Robotics, aided by AI and machine learning, will shortly change all sectors.

8. FUTURE SCOPE

Further this work will be implemented using solar panel and the conditions will be evaluate and monitor under deep learning and machine learning techniques. Different types of algorithm will be checked and carried out for the testing the automation process for running the agricultural robot. By using this work, farmers burden will get reduce and the automation process will be carried in the future generation.

REFERENCES

Bansod, S. & Jaiswal, R. (2018). Arduino Based Water Irrigation System. *International Journal of Advance Research in Science and Engineering, 7*.

Barhate, D. (2018). Design & Manufacturing of Multipurpose Agricultural Robot. *International Journal for Scinetific Research & Development*, *6*(2).

Chandrakant, D. & Deshmukh, S. M. (2020). Solar Powered Multi-Function Agri-Robot. *International Research Journal of Engineering and Technology, 7*(6).

Golbol, F., & Olmez, H. (2021). Developing a Motion Controller for Autonomous Agricultural Robot. *2020 28th Signal Processing and Computer Applications Conference*.

Jerosheja, B. R., & Mythili, C. (2020). Solar Powered Automated Multi-Tasking Agricultural Robot. *International Conference on Innovative Trends in Information Technology*.

Kurdi, M. M., & Elzein, I. A. (2022). Trajectory and Motion for Agricultural Robot. *2022 International Conference on Decision Aid Sciences and Applications*. 10.1109/DASA54658.2022.9765282

McGlym, S., & Walters, D. (2019). Agricultural Robots: Future Trends for Autonomous Farming. *Journal of Emerging Technologies and Innovative Research, 6*(4).

Muthurasu, S. & Divya. V. (2019). Design and Implementation of Agrobot by Using IOT. *International Journal of Advance Research, Ideas and Innovations in Technology, 5*(2).

Prajith, A. S. & Nowfiya B.S. (2020). Automatic Agricultural Robot-Agrobot. 2020 IEEE Bangalore Humaitarian Technology Conference. IEEE.

Priyadharsini, M. & Indhumathi, U. (2019). Automatic Irrigation System Using Soil. *International Journal of Engineering Trends and Technology, 67*(3).

Ranjitha, B., & Nikhitha, M. N. (2019). Solar Powered Autonomous Multipurpose Agricultural Robot Using Bluetooth/Android App. In *3rd International Conference on Electronics, Communication and Aerospace Technology*. IEEE.

Ratnadip, S. & Patil, S. (2021). Design and Development of Multipurpose Solar Powered Agro-Robot. *International Journal of Research in Engineering and Science, 9*(8).

Santos, L. (2020). Deep Learning Applications in Agriculture: A Short Review. *2019 Fourth Iberian Robotics Conference*. 10.1007/978-3-030-35990-4_12

Saravanan, N. (2021). Agricultural Robot using Machine Learning and Computer Vision. *International Research Journal of Engineering and Technology, 8*(3).

Shamshiri, S. & Weltzien, C. (2018). Research and Development in Agricultural Robotics: A Perspective of Digital Farming. *International Journal of Agricultural and Biological Engineering, 11*(4).

Shwetal Ramesh Gund, V. P. (2015, July). A Review on Agricultural Robots. *International Journal of Advanced Research in Computer Engineering and Technology*, *4*(7).

Sneha, A., Amrita, E., & Abirami. (2015). Agricultural Robot for Automatic Ploughing and Seeding. *2015 IEEE Technologies Innovation in ICT for Agriculture and Rurla Development*. IEEE.

Tejaswani, G. & Afroz, B. (2018). Agriculture Robotics. *International Journal of Engineering and Computer Science, 7*(3).

Xie, B., & Liu, J. (2021). Research Progress on Autonomous Navigation Technology of Agricultural Robot. *2021 IEEE 11th Annual International Conference on Cyber technology in Automation, Control and Intelligent Systems*. IEEE .

Zole, K., & Gedam, S. (2018). Agricultural Robot. *International Research Journal of Engineering and Technology, 5*(2).

Chapter 3
Applications of Deep Learning and Machine Learning in Smart Agriculture:
A Survey

Amrit pal Kaur
Manipal University Jaipur, India

Devershi Pallavi Bhatt
Manipal University Jaipur, India

Linesh Raja
https://orcid.org/0000-0002-3663-1184
Manipal University Jaipur, India

ABSTRACT

Machine learning (ML) and deep learning can be used in the smartest way possible to improve productivity in agriculture. The Food and Agriculture Organization's research shows that the crop's production is rising. One of the finest methods to monitor agricultural yield is through smart agriculture. Applications of ML and deep learning help to discover and resolve problems that crop growth encounters. In agriculture, the production of crops can be enhanced by applying machine learning and deep learning methodologies. These methods demonstrate the rapid advancement of artificial intelligence in the agriculture sector. The idea of "smart farming" keeps an eye on all processes, disease prediction, and agricultural pests. ML is used to extract meaningful information from huge datasets. Deep learning evaluates structural characteristics, meteorological data, and climatic factors to help anticipate agricultural diseases using practical and economical techniques. The deep-learning techniques enhance agricultural research's capacity to sense the overall classification of agriculture.

DOI: 10.4018/978-1-6684-9975-7.ch003

INTRODUCTION

Crop cultivation and raising livestock are two essential components of farming or agriculture. A nation's economy benefits immensely from it. A vast variety of food products and raw materials are produced by agriculture. A vast range of daily necessities are produced from raw materials utilized by industries like cotton and jute. Food and the raw materials required to generate commercial items are both produced in part by agriculture. In agriculture, standard techniques were used to cultivate crops. The majority of people worldwide use traditional or conventional farming. These procedures are time-consuming and labour-intensive because they are imprecise. Robotics, electronics, sensor, and automation technologies are all included in the field of precision agriculture as a means of using technology. Reduced workloads, increased profitability, and improved decision-making are the objectives of this technology. An agricultural management system that provides a thorough approach to regulating crop and soil variability across both spatial and temporal in order to maximize profitability, increase production, and improve the quality of products. Precision agriculture is a useful method for improving production. In comparison to low-value enterprise farms, high-value enterprise farms had a greater adoption rate for precision agriculture (Paustian, M., & Theuvsen, L., 2017). The country and geographical regions have an impact on the adoption rate of precision agriculture. Precision agriculture adoption among farmers in the hills zone is lower than those in the valley (Groher, Tanja, et al, 2020). Because such substantial expenditures are required, adoption rates might differ. Thus, finding a technique to lower the cost of the devices is necessary to enable precision agriculture on all types of fields. Modern technologies that enable precision agriculture include the Internet of Things, data mining, AI, and data science. The Internet of Things is a system of networked computing devices, such as sensors and smart devices, that can exchange data and interconnect with each other (Abdmeziem, M. R. et al, 2016). In order to anticipate crop health, agronomic apps employ wireless sensor networks to remotely monitor environmental and soil factors. To predict the timing of irrigation applications in agricultural areas, wireless networks of sensors may be utilized. Wireless sensor networks gather data regarding pressure, humidity, temperature, soil moisture, salinity, and conductivity, among other external factors. Machine learning has made agricultural applications easy to use and effective. Machine learning includes three stages: data gathering, model construction, and generalization (Akhter, R., & Sofi, S. A.,2022). In a large number of cases where human expertise is insufficient, machine learning algorithms are utilized to handle complicated issues. The prediction of soil properties like organic carbon and moisture level, crop production, diseases and weed detection in crops, and crop types recognition, may all be performed in agriculture using machine learning. Depending on the network architecture used, deep learning improves on traditional machine learning by adding more complexity to the model and modifying the input with different functions that enable structured data representation through many layers of abstraction. Automatic feature extraction from the source data is one of deep learning's main advantages (Sharma, A., et al., 2020). The Deep learning approach, which uses an agricultural field's homogeneous characteristics to locate far-off, badly obstructed, and unidentified items, makes use of the ability to distinguish unknown things like anomalies rather than just a collection of already-existing objects. Deep learning, machine learning, and the Internet of Things (IoT) are some of the data-focused technologies that are very beneficial for understanding and providing insightful data. In a variety of agricultural operations, these cutting-edge technologies are used to provide information about crop growth and assist in decision-making, such as choosing the best crop for a particular site and spotting elements that could harm the crops, such as weeds, insects, and crop diseases (Kamilaris, A., et al 2018).

Literature Review

There has been a lot of recent research on the use of machine learning techniques in the field of smart farming. In (Agana, N. A., & Homaifar, 2017), the author investigates the problem of drought prediction using deep learning algorithms. In order to forecast drought over the long term, it is suggested to use a Deep Belief Network with two Restricted Boltzmann Machines and lagged Standardized Streamflow Index (SSI) data as inputs. The study compares the efficacy of the proposed technique to that of conventional methods, such as Support Vector Regression (SVR) and Multilayer Perceptron (MLP), for forecasting the various time scales of drought conditions. The findings showed that the recommended strategy worked better than conventional methods. In order to predict agricultural output over time, they proposed a variety of techniques (Balakrishnan, N., & Muthukumarasamy, G., 2016). The Naive Bayes methodology and the SVM technique are contrasted with this method. The two metrics used to predict the result are accuracy and classification error. (Muangprathub J., et al. 2019) employed Internet of Things (IoT) technology to control environmental issues in agricultural areas. A web application, a mobile application, and physical components make up its three components. Regarding their distinctive designs and contributions to a range of disciplines, such as agriculture, the environment, and energy management, using IoT applications and machine learning techniques, some IoT-based machine learning models have been examined in (Din, I. U., 2019). In order to coordinate data-gathering efforts to solve common PA and PV issues, the mySense system was proposed in (Morais R. et al. 2019). The system is composed of four layers: cloud services, crop field and sensor networks, sensor and sensor node networks, and front-end application support. (Mahbub M. 2020) devised a sophisticated agricultural monitoring system to keep track of soil humidity and temperature. The system analyzes the observed data automatically and performs the required action based on readings of soil temperature and humidity. In (Moore, J., 2006), the author discusses a number of machine learning algorithms for forecasting flood risk. The findings are split into three categories: normal, abnormal, and high-risk floods. Numerous research show that pre-processing flood data with AI algorithms can enhance the result. The accuracy of classification techniques has increased as a result of these techniques. Neural network engineering normally produces respectable results in many applications, but our research has shown that random forest work produces the greatest results when compared to benchmark models. The author (Kishore KVK et al., 2020) explains how to use the group model to anticipate water levels in relation to the severity of the flood. Utilizing the most recent developments in the Internet of Things (IoT) and machine learning, this approach automates the analysis of flood data, which may be useful for preventing natural disasters. Findings from the research suggest that ensemble learning offers a more precise method for predicting flood severity levels. The authors (Ashok T., Varma PS., 2020) described farming techniques based on integrated systems, Internet systems, and wireless sensor networks for agricultural areas and animal farms. The study examines electrical circuit systems for systems, smart remote surveillance systems for computers, smartphones, and other devices. includes some suggestions towards the conclusion. A broad summary of the various methods for crop selection, crop sowing, threat detection, system monitoring, and finally production is provided by the authors in (Khalaf, Mohammed, et al. 2018). This project concentrated on the entire system that uses image processing, the Internet of Things, machine learning, and artificial intelligence. The authors in (Khalaf M. (2020) make their case for the optimum use of water to maximize production by evaluating the relationship between several physical parameters, such as soil temperature, soil moisture, air temperature, UV density, and air humidity, and plant development. Utilizing machine learning techniques to predict soil moisture helps conserve water resources.

Sujawat (2021) proposed using solar-powered Internet of Things sensor nodes to operate and monitor the agricultural sectors. Agriculture will also be monitored and regulated for activities including crop management, crop harvesting, animal control, pesticide distribution, and moisture and temperature measurement technology. According to Sharma et al. (2019) sensors may be able to identify field variables including temperature, humidity, and agricultural fertility. The sensing value is confirmed before being communicated to the Wi-Fi, and the Wi-Fi module's verified data is then sent to the farmer's laptop or mobile device through the cloud. Farmers can also receive SMS notifications if their fields need attention. In order to regulate the amount of water in a Microcontroller node, an algorithm including temperature, humidity, and fertility thresholds has been developed. The engine may be controlled by farmers from any location in the world. To analyze weather patterns and identify the optimum external conditions for generating high crop yields in a specific agricultural area, Suchithra et al. (2018) offered instructions on how to create wireless agricultural environmental sensor nodes. The major focus of this study is the literature on developing wireless agricultural environmental sensor nodes to monitor weather patterns and identify the perfect external circumstances for high crop yields in a specific agricultural region. The industry of agriculture and food production has lately shifted its attention to WSN with the goal of increasing production and establishing standards for yields from agriculture using this effective modern technology. Wireless sensor networks (WSNs) have received a lot of interest recentlyThe current analysis, according to Joshi et al. (2017), is a complete evaluation of the use of multiple algorithms based on machine learning for sensor data analysis in the agricultural system. It features a case study of a smart farm prototype powered by the Internet of Things that integrates food, energy, and water (FEW) systems. Mekonnen et al. (2019) claim that by examining a variety of meteorological variables, such as precipitation, temperature, and wetness, soil pH, soil type, and historical plant crop records, a machine learning technique is able to predict the best crop yield of a particular location. Sangeeta, (2020) recommended farmers monitor the fertility of the soil via data extraction analyses. The approach places a strong emphasis on soil quality monitoring in order to determine which crops are acceptable for cultivation based on the type of soil and to maximize crop yield using the right fertilizer. Ghadge et al. (2018) addressed the wide range of applications for artificial intelligence. Agricultural diseases may be greatly reduced thanks to artificial intelligence, which has the capacity to comprehend issues, identify the causes of them, and come up with the best treatments. The paper provides a brief overview of the use of artificial intelligence in agriculture, the existing agricultural techniques, and the many approaches accessible to identify plant diseases.

In this research, three important crops—such as rice, wheat, and sugarcane—have been taken into account. The macronutrient and micronutrient content of the soil, which is again a parametric representation of many climatic factors including rain, humidity, temperature, sunshine, and pH content of the soil, is typically a determinant of crop development. In this study, the micronutrients and meteorological data were established and related, and to classify different kinds of crops according to the micronutrients, machine learning, and decision tree algorithms were employed. For making predictions, a variety of methods have been employed, including curve fitting and data analysis using Python 3.9.0 (Dash, R et al, 2021). It is possible to calculate the GDD Growing Degree Days using the base temperature of a particular crop. This work's main objective is to offer straightforward, mathematically valid equations for calculating the GDD's base temperature. Using temperature data for snap beans, sweet corn, and cowpeas, mathematical formulas are suggested, shown, and put to the test. These new mathematical formulae can compute the base temperature more quickly and precisely than earlier techniques. These formulas may be used to predict the GDD base temperature for every crop at any developmental stage (Yang, S. et al., 1995). This study provides a formula to calculate the GDD of the crops. The formula provided in was used to the anticipated crop in order to assess the crop's GDD for the planned research. Using K-means and CNN models, weeds that are grown alongside soybeans may be found. Convolutional neural networks were utilized to categorize the weeds and soybeans, and K-means were employed to detect the characteristics of the photos. Additionally, it implies that by tweaking the CNN model, accuracy may be increased. The CNN model offers a reliable method for identifying weeds in agricultural areas. The photos and their augmentations are initially grouped when employed with K-means, and the CNN model then helps to properly identify the weed. According to the experimental findings, this approach with K-means pre-training obtained accuracy of 92.89%, outperforming convolutional neural networks with random initialization by 1.82% and two layer networks with no fine-tuning by 6.01%. Our results imply that adjusting parameters may increase identification accuracy (Tang, J. et al 2017).

Weed detection is done using currently available deep learning algorithms. This paper offers details on a variety of ML and Deep Learning methods that may be used to the identification of weeds. It focuses mostly on pre-trained models. This means that pre-trained models may be used to classify images and have a number of benefits. It also provides guidelines on how to handle datasets and prepare them for use in model creation. Several public datasets that are available on different websites can be used for this. It claims that techniques such as image scaling, data augmentation, and picture segmentation will provide accurate classifications, and pre-trained models also have a better propensity to boost accuracy. Since this research provides guidelines on how to implement deep learning techniques, the proposed strategy has chosen particular preprocessing steps such as Image Resizing and Data Augmentation before constructing the actual deep learning model to predict the weeds (Hasan, A. M., et al., 2021). In contrast, a novel approach is put forward in this research that blends deep learning with image processing technologies. First, veggies were identified and bounding boxes were created around them using a trained CenterNet model. The remaining green things that continued to fall out of the surrounding boxes were then classified as weeds. By concentrating just on detecting vegetables in this way, the model may avoid dealing with different weed species. Additionally, this approach may significantly minimize the complexity of weed detection as well as the amount of the training picture dataset, improving the performance and accuracy of weed identification. An image processing technique called colour index-based segmentation was used to separate the weeds from the backdrop. The selected color index was determined and evaluated in line with the

Bayesian classification error using Genetic Algorithms (GAs). In the field test, the trained CenterNet model achieved an F1 score of 0.953 with 95.6% accuracy and 95.0% recall (Jin, X, et al., 2021). A major amount of the crop's production and crop quality is destroyed by pest infestation, making it challenging for farmers to identify agricultural insects. The disadvantage of traditional insect identification is that accurate physical characteristic-based identification of insects requires taxonomists with advanced degrees. Tests were undertaken to identify 24 insects from the Wang and Xie dataset using visual cues and a number of methods, including neural networks, support vector machines, k-nearest neighbors, naive bayes, and convolutional neural network models. Utilizing 9-fold cross-validation improved the categorization models' efficacy. The best classification rates (91.5 and 90%, respectively) were achieved with the CNN model. When compared to cutting-edge classification approaches, the findings showed a noticeably higher classification accuracy and calculation speed. This work used both the core CNN model and the same dataset as other studies' (Kasinathan, T., et al., 2021). A separate dataset from the Kaggle platform called the Pests' dataset was therefore used in the recommended model. In this collection, there are nine different classes of insects. Each picture was taken from a different location. The proposed model was selected for this dataset because it has been trained on photos from various locations, which gives it more information to assess the images and distinguish between them. This study's main objective is to pinpoint the critical elements that influence plant development. This essay discusses the notion of reducing agricultural hazards and fostering intelligent farming. The AI-based wireless sensor will increase the standard for intelligent agriculture, although agriculture has always improved. This study's goal is to create machine learning techniques that use visual processing to improve prediction states. The paper's main goal is to identify and treat illnesses affecting cotton leaves. The study has several different components, including soil sensing, server-based remote monitoring, moisture and temperature sensing, and leaf disease identification. Insects and diseases frequently cause plant ailments that, if not promptly treated, reduce production. This study presents a method for maintaining soil quality monitoring and guarding against diseases of cotton leaves. The regression methodology of artificial intelligence is suggested as a way to identify and classify leaf illnesses (Altalak, M., et al., 2022). Farmers would be informed of the ailment if it was discovered using the Android app. Along with the chemical level in a container, the Android app also allows for the display of soil parameter values like temperature, humidity, and wetness. The relay may be activated or deactivated using the Android app as necessary to regulate the motor and chemical spray system. The SVM algorithm improves farmers' crop yields in the recommended system, demonstrating its usefulness in the diagnosis and control of diseases. (Murugamani, C.,et al, 2022) In order to predict drought and crop yield, this research suggests an IoT-based smart farming system with the successful prediction method WPART. WPART is built on machine learning techniques. Farmers and agricultural executives place a high value on crop production and drought forecasts, which considerably assist nations throughout the world that are affected by agriculture. Drought early warning plays a critical role in reducing the impacts of drought on agricultural yield, and drought prediction research attempts to increase predictability skill by fully using sources of predictability and enhancing our understanding of the physical mechanism of drought. In this study, a smart strategy for forecasting agricultural productivity and drought is provided. It combines a wrapper feature selection approach with a PART classification algorithm (Rezk, N. G et al 2021). The suggested approach is estimated using five datasets. Comparing the projected approach to the current ones, the findings showed that it is more reliable, accurate, and exact at classifying and predicting agricultural productivity and drought. According to the findings, the suggested technique is the most reliable for predicting droughts and agricultural yield for crops including sugarcane, soy, jowar, and bajra. When compared to the current supreme standard techniques, the WPART approach achieves the highest levels of accuracy; for the five datasets used for predicting droughts and measure crop production, it achieves accuracy levels of 92.51%, 96.77%, 98.04%, 96.12%, and 98.15%, respectively.

This review makes a contribution to the availability of cutting-edge deep learning-based computer vision technologies that may help farmers with tasks ranging from land preparation through harvesting. The examination of recent works in computer vision was divided into (a) seed quality analysis, (b) soil analysis, (c) irrigation water management, (d) plant health analysis, (e) weed management, (f) livestock management, and (g) yield estimate in this research. In addition, the study examines current developments in computer vision, including generative adversarial networks (GAN), vision transformers (ViT), and other well-liked deep learning architectures. This study also identifies the difficulties in immediately putting the recommendations into practice in the fields of farmers. Overall, the results show that convolutional neural networks are the foundation of contemporary computer vision methods, and that the different designs of these networks offer very accurate and precise solutions for a variety of agricultural tasks (Dhanya, V. G., et al, 2022). However, the key to the computer vision approach's success is using a high-quality dataset to create the model and offering real-time answers.Long Short-Term Memory (LSTM)-based neural networks are suggested as an alternative strategy in this research to overcome this issue (Sami, M et al 2022). The suggested method, which substitutes a neural sensor for a physical sensor in a smart irrigation system, is put to the test. A number of physical sensors are part of the Smart Irrigation System (SIS), which transmits data on temperature, humidity, and soil moisture to determine transpiration in a given area. The real-world data were obtained from an agricultural field in a lemon orchard close to the Pakistani province of Ghadap Sindh. The LM35 sensor is used to measure temperature, the DHT-22 sensor is used to measure humidity, and we created a specially developed sensor in our lab to measure moisture levels. The experiment's findings demonstrate that the suggested deep learning-based neural sensor accurately predicts real-time readings, particularly temperature measurements. The readings for moisture and humidity are likewise within acceptable bounds. Our findings show that a neural network, referred to as a neural sensor here, may be used to augment the functionality of a physical sensor placed in an agricultural field in order to increase the accuracy of smart irrigation systems (Sami, M et al 2022).

Applications of Machine Learning and Deep Learning in Smart Agriculture

The applications of machine learning and deep learning in smart agriculture are covered in this section of the book's content. To duplicate human learning processes, learn new knowledge, continually improve performance, and mature to a certain degree, robots may be programmed using ML and DL, two cutting-edge technologies. In recent years, these strategies have achieved great success in algorithms, theories, and implementations when paired with other agricultural strategies to reduce crop expenses and increase production. Some application of machine learning in smart agriculture are discussed below. Figure. 1 illustrates different applications of machine learning and deep learning in smart agriculture.

Crop Management

1. Analysis of seed quality

Farmers will receive the right quantity and quality of seeds at the right time due to the commercial seed industry. To distinguish between low-quality and high-quality seeds, it is labor-intensive, time-consuming, and requires sophisticated equipment and infrastructure (Kannur et al., 2011). The testing of seeds for their quality may really accelerate with the use of computer vision technology, which can extract morphological data from different seed batches and grade it in line with the quality criteria established globally (Bao and Bambil, 2021). The various seed testing modules will probably concentrate on the seed's physical purity, genetic purity, health, vigor, and patterns of deterioration. In general, this may cover the physical or visually attributable characters like the seed length, shape, and size, as well as any visual impairments and the presence of foreign bodies, which can be captured by sophisticated computer vision technology (Granitto et al., 2005). Computer vision based on deep learning has substantially improved on classical computer vision's performance difficulties, leading to a broader

Figure 1. Applications of machine learning in agriculture

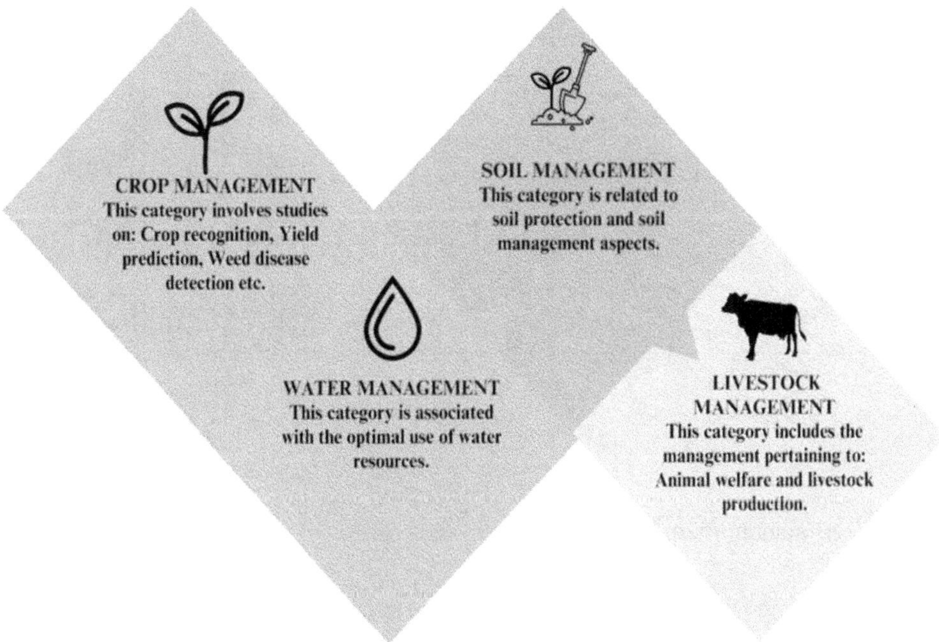

usage for seed variety detection. Figure. 2 illustrates how computer vision is used to assess the quality of seeds. To increase the accuracy of these methods, spectral imaging techniques are frequently used with them (Qiu et al., 2018). In a study by (Zhu et al. 2019), it was demonstrated that combining spectroscopy with machine learning - CNN models - was successful in identifying the seed types. The CNN and ResNet models' feature reconstructions, which were used to identify the cotton seeds, showed an accuracy was over 80 percent. When used to sort maize and sunflower seeds, SeedSortNet, which was created using CNN models for computer vision, was shown to be accurate to 97.33% and 99.56% (Li et al., 2021). It was successful in predicting viability in 90% of the cases for naturally aging seeds. In order to automatically identify different chickpea types. Models were developed using visible-spectrum pictures of the seeds by (Taheri-Garavand et al. 2021). The identification was performed using a modified VGG16 model. Selecting high-quality seeds is crucial for increasing output in the breeding business, hence (Zhao et al. 2021) employed seven distinct computer vision models to correctly detect and identify surface faults. For the MobileNet-V2 model, the soybean dataset achieved excellent detection accuracy. Numerous studies of this nature have been carried out by various researchers, and the seed industry is substantially benefited by state-of-the-art computer vision models, acquiring a higher degree of automation capabilities.

Figure 2. Seed quality analysis

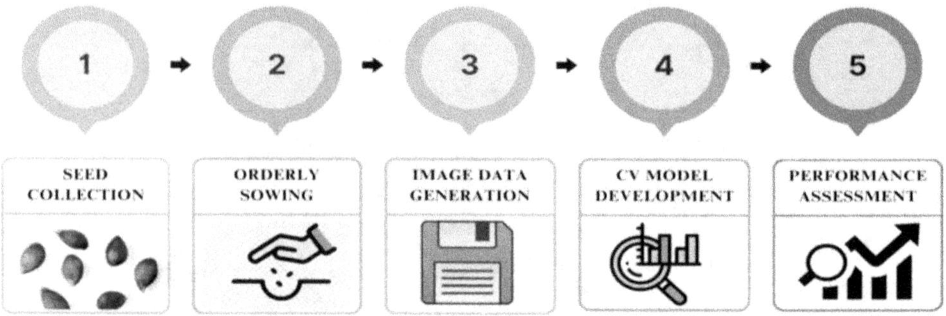

2. Weed management

Weeds are one of the main aspects that harm agricultural productivity. The focus on improving agricultural productivity is evidently resulting in an increase in the number of chemicals discharged into the environment with the intention of reducing weed growth. But in order to enhance production, resources must be used as effectively as possible, and this can only be achieved by carefully spraying weeds. Traditional robotic weeders don't use crop identification technology to carry out their weed operations; instead, they typically work by recognizing crop row patterns. Large weed populations and densities may hide the row pattern, reducing the effectiveness of the weeders. As precise weed spraying depends on the proper identification and positioning of weeds, computer vision technologies save the day at this point by precisely recognizing the items. Recent studies have examined the viability of using computer vision technology for agronomic plant species categorization at the field level, such as the distinction between crops and weeds and other undesirable species (Sau and Ucchesu, 2019; Sau et al., 2018; Subeesh et al., 2022). A thorough implementation of the same has been recommended for automatic plant species identification based on the leaf recognition pattern (Putzu et al., 2016), as well as the botanical characterisation of germplasm (Lo Bianco et al., 2017), for the goal of safeguarding and cataloguing plant species. The main methods for weed detection at the field level are deep learning computer vision technology and traditional image processing. It is crucial to separate out the different components, such as color, form, texture, etc., when utilizing conventional computer vision techniques and to integrate them with machine learning methods, such the SVM. However, due to their improved ability to express image data and avoid the drawbacks of traditional methods of feature extraction, deep learning algorithms have become more advantageous for extracting the multidimensional and multiscale spatial and semantic feature information of weeds through AlexNet, VGGNet, ResNet, etc. as computing power has increased. The application of deep learning in the agronomic classification of plant species has picked up steam since the introduction of CNN and AlexNet (Krizhevsky et al., 2012). Using around 1900 images of the same, (Hall et al. 2015) categorized the leaves of 32 different types of crops and weeds using CNN architecture. According to (Kussul et al. 2017), CNN architecture might be used to classify and separate weeds from different varieties of wheat, sugarbeet, corn, soybean, sunflower, etc. A modified version of VGG16 was recommended by (Mortensen et al. 2016) for the categorization of weeds, grass, barley, and oil crops.

Water Management

It takes a lot of work to manage irrigation water in agricultural production, but it's essential for preserving the hydrological, climatological, and agronomic balance. In order to learn more about the biophysical processes involved in water absorption via the soil's root zone and transpiration through the plant canopy, a number of research have been conducted (Kushwaha et al., 2021; Elbeltagi et al 2022). A successful irrigation plan must take into account the precise amount of water a crop requires (Kushwaha et al., 2016; Vishwakarma et al., 2022). Consequently, it is essential to adopt automated methods for managing agricultural output, watering plants, and evaluating productivity, as well as computer vision technology. Wang et al. Convolutional Neural Networks (CNNs) were used in the detection and monitoring of center pivot irrigation systems by (Zhang et al., 2018) to allocate irrigation water. Similar to this, (Chang and Lin, 2018) created a small, intelligent agricultural equipment that uses computer vision and multitasking to autonomously weed and hydrate planted ground. In order to water and weed while maintaining an average herbicidal rate of 90% and a deep soil moisture level of 80%, the system identifies the plants and weeds in real-time. Because it enables resource use in its totality as well as multitasking integration, this technique has a lot of potential. Kamyshova et al.'s paper (Kamyshova et al., 2022) offered an algorithm-based method for producing a maize irrigation map in order to optimize the watering process of crops using a phytoindicator system in low latency mode. The system, which consists of eight IP cameras connected to a laptop through a DVR, may be put on a pivot irrigation system. The algorithm is broken down into three phases. Utilizing the ExGR (integrated excess green and excess red difference) index during the preprocessing step of the picture. The categorization step involves implementing the strategy that the research selected in accordance with the operating circumstances of the system. The last step determines the rate of watering plants in the current sector of the sprinkler site using a neural network trained using the Resilient Propagation approach. The development phases and accuracy of plant identification were both up to 93 percent. All technological fields are currently using inexpensive cameras, but agricultural applications are notable. Using pictures to gather pertinent data on the growth of horticultural crops, the moisture content of the soil may be properly measured to enable the exact scheduling of irrigation (Koech and Langat, 2018).

Soil Management

According to Kushaha et al. (2022; Suchithra and Pai, 2020), the main objective of soil management in agriculture is improving and preserving dynamic soil features. This will increase crop yield. Traditionally, soil samples are collected and brought to a lab for drying out, squeezing, and sieving before being utilized in the examination of soil texture. The most common laboratory analytical technique for soils with a coarse texture or sand content is sieving, whereas a hydrometer or pipette approach based on the sedimentation hypothesis is utilized for soils with finer textured particles (Kushwaha et al., 2022; Sudarsan et al., 2016). Computer vision-based techniques for image analysis have drawn a lot of interest from a variety of industries, including soil science, due to recent improvements in image processing capacity and the development of image capture (such as camera) systems. Using basic computer programs, this technique employs cameras to capture soil images (static or dynamic), which are subsequently categorized and classified shown in (Figure. 3). When textural patterns are matched, it may be possible to determine the size of the soil particles, for example, directly from the image. Several studies have tested a variety of image analysis-based computer vision methods. Entropy and angular moment-based textural categorization were employed by (Haralick et al. 1973) in an effort to categorize images collected from an aerial or satellite perspective. Since then, a number of remote sensing applications have made use of the gray level co-occurrence matrix (GLCM) and its analogs (Dell'Acqua and Gamba, 2003; Kuplich et al., 2005). The highest resolution satellite, however, is only capable of providing a maximum resolution of 10 m/square pixel, which is insufficient to grasp soil particle sizes. Three 1-dimensional (1D) convolutional neural networks—the LucasCNN, the LucasResNet, and the LucasCoordConv—were put into practice (Riese and Keller 2019). Additionally, the study modifies two existing 1D CNN algorithms for the given classification task and assesses how well they perform in comparison to a random forest classifier. Thus, the study makes use of the open-source LUCAS topsoil dataset. The best accurate classifier is found to be the CNN approach with the least amount of depth. The results of the LucasCoordConv are the best in terms of average accuracy.

Figure 3. Soil management process

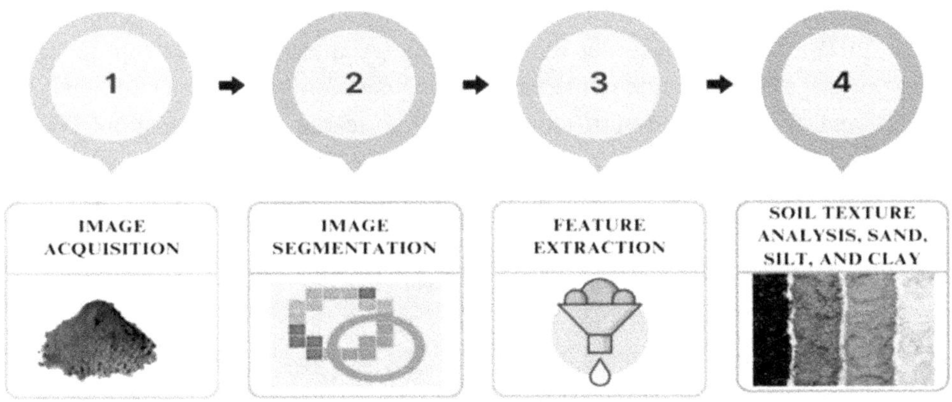

Livestock Management

In order to ensure the health and productivity of each individual animal, precision livestock farming (PLF) heavily relies on computer vision techniques. Systems for monitoring livestock in real time give farmers information they may use to make informed decisions shown in (Figure. 4). Non-invasive computer vision technology for the detection of cattle behavior has been the subject of several research recently (Bello et al., 2021; Shen et al., 2020; Kumar et al., 2017; Qiao et al., 2019). To recognize cows in an unconstrained barn, (Xiao et al., 2022) used a modified Mask-RCNN model and trained a fusion of Mask-RCNN and SVM. A collection of 1,553 photos was used by Hansen et al. (2018) to train a CNN to detect pigs by their faces. The VGG-face model employed in this investigation has a 96.7% accuracy rate. Unmanned aerial vehicles (UAVs) were used in several research to collect data in order to precisely locate and count the livestock (Andrew et al., 2019; Rivas et al., 2018; Chamoso et al., 2014; Rahnemoonfar et al., 2019). These challenges involving animal detection and counting often use either the development of bounding boxes for animal detection or the generation of CNN-based probability heat maps on the location of the animals. Jiang et al. (2019) introduced the 'FLYOLOv3' (FilterLayer YOLOv3) Yolo model, which is an enhanced Yolo model based on Filter layer and ensures reliable recognition of important dairy cow sections. This method outperformed the Yolov3 and Faster-RCNN algorithms in terms of performance. Dairy cow health and production are tightly correlated with daily activity patterns, food consumption, and rumination (Weary et al., 2009;Huzzey et al., 2007). Recent research highlights the gradual replacement of time-lapse video capture and direct observation with computer vision techniques. In order to separate specific pigs from a group and determine the pace at which they occupy the feeding area, Yang et al. (2018) employed a Faster-RCNN model. (Alameer et al. 2020) proposed a GoogleNet-based technique to increase the precision of feeding behavior analysis, detect, and eliminate non-nutritive visits (NNV) to the feeding location. With a 99.4% accuracy rate, dietary patterns could be detected with great precision. In the animal husbandry farm, CNN architectures have been found to be effective in detecting early cow sickness (Rony et al., 2021).

CONCLUSION

The adoption of deep learning, as well as machine learning over the last few years, has given traditional agricultural techniques the ability to be automated. The contribution of deep learning-based computer vision to many agricultural applications has been comprehensively covered in this study. The study focuses on six particular application areas, including analysis of seed quality, analysis of soil, water management, managing weeds, livestock management, and crop yield

Figure 4. Livestock management process

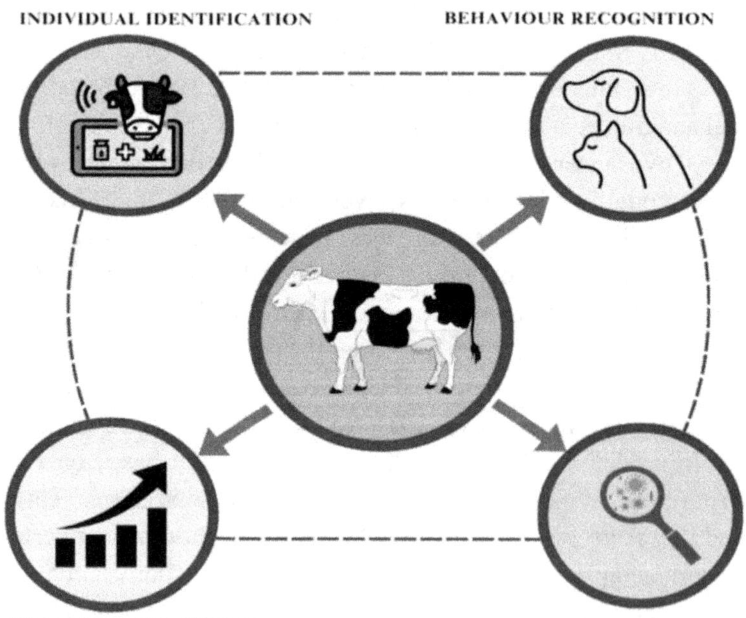

prediction. According to an analysis of deep learning's applications, particularly for the evaluation and scheduling of water resources, machine learning techniques are still being used in the water industry at a rapid rate and will have a significant impact on future water-related research as well as the broad range of application areas. Technologies based on deep learning have opened up a wide range of research and application possibilities that might revolutionize hydrological science and practice. Recent developments in image classification, object recognition, segmentation, and other methods for deep learning-aided image analysis have broadened their applications to various pre- and post-harvesting operations in agriculture.

REFERENCES

Abdmeziem, M. R., Tandjaoui, D., & Romdhani, I. (2016). Architecting the internet of things: state of the art. *Robots and Sensor Clouds*, 55-75.

Agana, N. A., & Homaifar, A. (2017, March). A deep learning based approach for long-term drought prediction. In *SoutheastCon 2017* (pp. 1–8). IEEE. doi:10.1109/SECON.2017.7925314

Akhter, R., & Sofi, S. A. (2022). Precision agriculture using IoT data analytics and machine learning. *Journal of King Saud University-Computer and Information Sciences*, *34*(8), 5602–5618. doi:10.1016/j.jksuci.2021.05.013

Alameer, A., Kyriazakis, I., Dalton, H. A., Miller, A. L., & Bacardit, J. (2020). Automatic recognition of feeding and foraging behaviour in pigs using deep learning. *Biosystems Engineering*, *197*, 91–104. doi:10.1016/j.biosystemseng.2020.06.013

Altalak, M., Uddin, M. A., Alajmi, A., & Rizg, A. (2022). A Hybrid Approach for the Detection and Classification of Tomato Leaf Diseases. *Applied Sciences (Basel, Switzerland)*, *12*(16), 8182. doi:10.3390/app12168182

Andrew, W., Greatwood, C., & Burghardt, T. (2019, November). Aerial animal biometrics: Individual friesian cattle recovery and visual identification via an autonomous uav with onboard deep inference. In *2019 IEEE/RSJ International Conference on Intelligent Robots and Systems (IROS)* (pp. 237-243). IEEE. 10.1109/IROS40897.2019.8968555

Ashok, T., & Varma, P. S. (2020). Crop prediction based on environmental factors using machine learning ensemble algorithms. In *Intelligent Computing and Innovation on Data Science* (pp. 581–594). Springer. doi:10.1007/978-981-15-3284-9_63

Balakrishnan, N., & Muthukumarasamy, G. (2016). Crop production-ensemble machine learning model for prediction. *International Journal of Computer Science and Software Engineering*, *5*(7), 148.

Bao, F., & Bambil, D. (2021). Applicability of computer vision in seed identification: Deep learning, random forest, and support vector machine classification algorithms. *Acta Botanica Brasílica*, *35*(1), 17–21. doi:10.1590/0102-33062020abb0361

Bello, R. W., Mohamed, A. S. A., & Talib, A. Z. (2021). Contour extraction of individual cattle from an image using enhanced Mask R-CNN instance segmentation method. *IEEE Access : Practical Innovations, Open Solutions*, *9*, 56984–57000. doi:10.1109/ACCESS.2021.3072636

Chamoso, P., Raveane, W., Parra, V., & González, A. (2014). UAVs applied to the counting and monitoring of animals. In *Ambient Intelligence-Software and Applications: 5th International Symposium on Ambient Intelligence* (pp. 71-80). Springer International Publishing. 10.1007/978-3-319-07596-9_8

Chang, C. L., & Lin, K. M. (2018). Smart agricultural machine with a computer vision-based weeding and variable-rate irrigation scheme. *Robotics (Basel, Switzerland)*, *7*(3), 38. doi:10.3390/robotics7030038

Dash, R., Dash, D. K., & Biswal, G. C. (2021). Classification of crop based on macronutrients and weather data using machine learning techniques. *Results in Engineering*, *9*, 100203. doi:10.1016/j.rineng.2021.100203

Dell'Acqua, F., & Gamba, P. (2003). Texture-based characterization of urban environments on satellite SAR images. *IEEE Transactions on Geoscience and Remote Sensing*, *41*(1), 153–159. doi:10.1109/TGRS.2002.807754

Dhanya, V. G., Subeesh, A., Kushwaha, N. L., Vishwakarma, D. K., Kumar, T. N., Ritika, G., & Singh, A. N. (2022). *Deep learning based computer vision approaches for smart agricultural applications*. Artificial Intelligence in Agriculture. doi:10.1016/j.aiia.2022.09.007

Din, I. U., Guizani, M., Rodrigues, J. J. P. C., Hassan, S., & Korotaev, V. V.I. U. (2019). Machine learning in the Internet of Things: Designed techniques for smart cities. *Future Generation Computer Systems*, *100*, 826–843. doi:10.1016/j.future.2019.04.017

Elbeltagi, A., Kushwaha, N. L., Rajput, J., Vishwakarma, D. K., Kulimushi, L. C., Kumar, M., Zhang, J., Pande, C. B., Choudhari, P., Meshram, S. G., Pandey, K., Sihag, P., Kumar, N., & Abd-Elaty, I. (2022). Modelling daily reference evapotranspiration based on stacking hybridization of ANN with meta-heuristic algorithms under diverse agro-climatic conditions. *Stochastic Environmental Research and Risk Assessment*, *36*(10), 3311–3334. doi:10.100700477-022-02196-0

Ghadge, R., Kulkarni, J., More, P., Nene, S., & Priya, R. L. (2018). Prediction of crop yield using machine learning. *Int. Res. J. Eng. Technol., 5*.

Granitto, P. M., Verdes, P. F., & Ceccatto, H. A. (2005). Large-scale investigation of weed seed identification by machine vision. *Computers and Electronics in Agriculture*, *47*(1), 15–24. doi:10.1016/j.compag.2004.10.003

Groher, T., Heitkämper, K., Walter, A., Liebisch, F., & Umstätter, C. (2020). Status quo of adoption of precision agriculture enabling technologies in Swiss plant production. *Precision Agriculture*, *21*(6), 1327–1350. doi:10.100711119-020-09723-5

Hall, D., McCool, C., Dayoub, F., Sunderhauf, N., & Upcroft, B. (2015, January). *Evaluation of features for leaf classification in challenging conditions. In 2015 IEEE Winter conference on applications of computer vision*. IEEE.

Hansen, M. F., Smith, M. L., Smith, L. N., Salter, M. G., Baxter, E. M., Farish, M., & Grieve, B. (2018). Towards on-farm pig face recognition using convolutional neural networks. *Computers in Industry*, *98*, 145–152. doi:10.1016/j.compind.2018.02.016

Haralick, R. M., Shanmugam, K., & Dinstein, I. H. (1973). Textural features for image classification. *IEEE Transactions on Systems, Man, and Cybernetics*, *SMC-3*(6), 610–621. doi:10.1109/TSMC.1973.4309314

Hasan, A. M., Sohel, F., Diepeveen, D., Laga, H., & Jones, M. G. (2021). A survey of deep learning techniques for weed detection from images. *Computers and Electronics in Agriculture*, *184*, 106067. doi:10.1016/j.compag.2021.106067

Huzzey, J. M., Veira, D. M., Weary, D. M., & Von Keyserlingk, M. A. G. (2007). Prepartum behavior and dry matter intake identify dairy cows at risk for metritis. *Journal of Dairy Science*, *90*(7), 3220–3233. doi:10.3168/jds.2006-807 PMID:17582105

Jiang, B., Wu, Q., Yin, X., Wu, D., Song, H., & He, D. (2019). FLYOLOv3 deep learning for key parts of dairy cow body detection. *Computers and Electronics in Agriculture*, *166*, 104982. doi:10.1016/j.compag.2019.104982

Jin, X., Che, J., & Chen, Y. (2021). Weed identification using deep learning and image processing in vegetable plantation. *IEEE Access : Practical Innovations, Open Solutions, 9*, 10940–10950. doi:10.1109/ACCESS.2021.3050296

Joshi, P. P., Kanade, S. S., & Joshi, S. P. (2017). Wireless sensor network and monitoring of crop field. *IOSR Journal of Electronics and Communication Engineering, 12*(1), 23-28.

Kamilaris, A., & Prenafeta-Boldú, F. X. (2018). Deep learning in agriculture: A survey. *Computers and Electronics in Agriculture, 147*, 70–90. doi:10.1016/j.compag.2018.02.016

Kamyshova, G., Osipov, A., Gataullin, S., Korchagin, S., Ignar, S., Gataullin, T., Terekhova, N., & Suvorov, S. (2022). Artificial neural networks and computer vision's-based phytoindication systems for variable rate irrigation improving. *IEEE Access : Practical Innovations, Open Solutions, 10*, 8577–8589. doi:10.1109/ACCESS.2022.3143524

Kannur, A., Kannur, A., & Rajpurohit, V. S. (2011). Classification and grading of bulk seeds using artificial neural network. *Jisuanji Yingyong, 3*, 62–73.

Kasinathan, T., Singaraju, D., & Uyyala, S. R. (2021). Insect classification and detection in field crops using modern machine learning techniques. *Information Processing in Agriculture, 8*(3), 446–457. doi:10.1016/j.inpa.2020.09.006

Khalaf, M., Alaskar, H., Hussain, A. J., Baker, T., Maamar, Z., Buyya, R., Liatsis, P., Khan, W., Tawfik, H., & Al-Jumeily, D. (2020). IoT-enabled flood severity prediction via ensemble machine learning models. *IEEE Access : Practical Innovations, Open Solutions, 8*, 70375–70386. doi:10.1109/ACCESS.2020.2986090

Khalaf, M., Hussain, A. J., Al-Jumeily, D., Baker, T., Keight, R., Lisboa, P., ... Al Kafri, A. S. (2018). A data science methodology based on machine learning algorithms for flood severity prediction. In *2018 IEEE Congress on Evolutionary Computation (CEC)* (pp. 1–8). IEEE 10.1109/CEC.2018.8477904

Kishore, K. V. K., Kumar, B. Y., & Venkatramaphanikumar, S. (2020). Optimized water scheduling using IoT sensor data in smart farming. In *Smart Technologies in Data Science and Communication* (pp. 205–219). Springer. doi:10.1007/978-981-15-2407-3_26

Koech, R., & Langat, P. (2018). Improving irrigation water use efficiency: A review of advances, challenges and opportunities in the Australian context. *Water (Basel), 10*(12), 1771. doi:10.3390/w10121771

Krizhevsky, A., Sutskever, I., & Hinton, G. E. (2017). Imagenet classification with deep convolutional neural networks. *Communications of the ACM*, *60*(6), 84–90. doi:10.1145/3065386

Kumar, S., Singh, S. K., Singh, R., Singh, A. K., Kumar, S., Singh, S. K., & Singh, A. K. (2017). Recognition of cattle using face images. *Animal Biometrics: Techniques and Applications*, 79-110.

Kushwaha, N. L., Bhardwaj, A., & Verma, V. K. (2016). Hydrologic response of Takarla-Ballowal watershed in Shivalik foot-hills based on morphometric analysis using remote sensing and GIS. *J Indian Water Resour Soc*, *36*(1), 17–25.

Kushwaha, N. L., Elbeltagi, A., Mehan, S., Malik, A., & Yousuf, A. (2022). Comparative study on morphometric analysis and RUSLE-based approaches for micro-watershed prioritization using remote sensing and GIS. *Arabian Journal of Geosciences*, *15*(7), 564. doi:10.100712517-022-09837-2

Kushwaha, N. L., Rajput, J., Elbeltagi, A., Elnaggar, A. Y., Sena, D. R., Vishwakarma, D. K., Mani, I., & Hussein, E. E. (2021). Data intelligence model and meta-heuristic algorithms-based pan evaporation modelling in two different agro-climatic zones: A case study from Northern India. *Atmosphere (Basel)*, *12*(12), 1654. doi:10.3390/atmos12121654

Kussul, N., Lavreniuk, M., Skakun, S., & Shelestov, A. (2017). Deep learning classification of land cover and crop types using remote sensing data. *IEEE Geoscience and Remote Sensing Letters*, *14*(5), 778–782. doi:10.1109/LGRS.2017.2681128

Li, C., Li, H., Liu, Z., Li, B., & Huang, Y. (2021). SeedSortNet: A rapid and highly effificient lightweight CNN based on visual attention for seed sorting. *PeerJ. Computer Science*, *7*, e639. doi:10.7717/peerj-cs.639 PMID:34435095

Lo Bianco, M., Grillo, O., Cañadas, E., Venora, G., & Bacchetta, G. (2017). Inter-and intraspecific diversity in Cistus L.(Cistaceae) seeds, analysed with computer vision techniques. *Plant Biology*, *19*(2), 183–190. doi:10.1111/plb.12529 PMID:27917577

Mahbub, M. (2020). A smart farming concept based on smart embedded electronics, internet of things and wireless sensor network. *Internet of Things*, *9*, 100161. doi:10.1016/j.iot.2020.100161

Mekonnen, Y., Namuduri, S., Burton, L., Sarwat, A., & Bhansali, S. (2019). Machine learning techniques in wireless sensor network based precision agriculture. *Journal of the Electrochemical Society*, *167*(3), 037522. doi:10.1149/2.0222003JES

Moore, J., Chase, J. S., & Ranganathan, P. (2006). *Weatherman: automated, online and predictive thermal mapping and management for data centers. In 2006 IEEE international conference on Autonomic Computing*. IEEE.

Morais, R., Silva, N., Mendes, J., Adão, T., Pádua, L., López-Riquelme, J. A., Pavón-Pulido, N., Sousa, J. J., & Peres, E. (2019). Mysense: A comprehensive data management environment to improve precision agriculture practices. *Computers and Electronics in Agriculture*, *162*, 882–894. doi:10.1016/j.compag.2019.05.028

Mortensen, A. K., Dyrmann, M., Karstoft, H., Jørgensen, R. N., & Gislum, R. (2016, June). Semantic segmentation of mixed crops using deep convolutional neural network. In CIGR-AgEng conference (pp. 26-29).

Muangprathub, J., Boonnam, N., Kajornkasirat, S., Lekbangpong, N., Wanichsombat, A., & Nillaor, P. (2019). IoT and agriculture data analysis for smart farm. *Computers and Electronics in Agriculture*, *156*, 467–474. doi:10.1016/j.compag.2018.12.011

Murugamani, C., Shitharth, S., Hemalatha, S., Kshirsagar, P. R., Riyazuddin, K., Naveed, Q. N., Islam, S., Mazher Ali, S. P., & Batu, A. (2022). Machine learning technique for precision agriculture applications in 5G-based internet of things. *Wireless Communications and Mobile Computing*, *2022*, 2022. doi:10.1155/2022/6534238

Paustian, M., & Theuvsen, L. (2017). Adoption of precision agriculture technologies by German crop farmers. *Precision Agriculture*, *18*(5), 701–716. doi:10.100711119-016-9482-5

Putzu, L., Di Ruberto, C., & Fenu, G. (2016). A mobile application for leaf detection in complex background using saliency maps. *Advanced Concepts for Intelligent Vision Systems: 17th International Conference, ACIVS 2016, Lecce, Italy, October 24-27, 2016 Proceedings*, *17*, 570–581.

Qiao, Y., Su, D., Kong, H., Sukkarieh, S., Lomax, S., & Clark, C. (2019). Individual cattle identification using a deep learning based framework. *IFAC-PapersOnLine*, *52*(30), 318–323. doi:10.1016/j.ifacol.2019.12.558

Qiu, Z., Chen, J., Zhao, Y., Zhu, S., He, Y., & Zhang, C. (2018). Variety identification of single rice seed using hyperspectral imaging combined with convolutional neural network. *Applied Sciences (Basel, Switzerland)*, *8*(2), 212. doi:10.3390/app8020212

Rahnemoonfar, M., Dobbs, D., Yari, M., & Starek, M. J. (2019). DisCountNet: Discriminating and counting network for real-time counting and localization of sparse objects in high-resolution UAV imagery. *Remote Sensing (Basel)*, *11*(9), 1128. doi:10.3390/rs11091128

Rezk, N. G., Hemdan, E. E. D., Attia, A. F., El-Sayed, A., & El-Rashidy, M. A. (2021). An efficient IoT based smart farming system using machine learning algorithms. *Multimedia Tools and Applications*, *80*(1), 773–797. doi:10.100711042-020-09740-6

Riese, F. M., & Keller, S. (2019). Soil texture classification with 1D convolutional neural networks based on hyperspectral data. *arXiv preprint arXiv:1901.04846.*

Rivas, A., Chamoso, P., González-Briones, A., & Corchado, J. M. (2018). Detection of cattle using drones and convolutional neural networks. *Sensors (Basel)*, *18*(7), 2048. doi:10.339018072048 PMID:29954080

Rony, M., Barai, D., & Hasan, Z. (2021, July). Cattle external disease classification using Deep Learning Techniques. In *2021 12th International Conference on Computing Communication and Networking Technologies (ICCCNT)* (pp. 1-7). IEEE. 10.1109/ICCCNT51525.2021.9579662

Sami, M., Khan, S. Q., Khurram, M., Farooq, M. U., Anjum, R., Aziz, S., Qureshi, R., & Sadak, F. (2022). A deep learning-based sensor modeling for smart irrigation system. *Agronomy (Basel)*, *12*(1), 212. doi:10.3390/agronomy12010212

Sangeeta, S. G. (2020). Design and implementation of crop yield prediction model in agriculture. *International Journal of Scientific & Technology Research*, *8*(1), 544–549.

Sau, S., Ucchesu, M., D'hallewin, G., & Bacchetta, G. (2019). Potential use of seed morpho-colourimetric analysis for Sardinian apple cultivar characterisation. *Computers and Electronics in Agriculture*, *162*, 373–379. doi:10.1016/j.compag.2019.04.027

Sau, S., Ucchesu, M., Dondini, L., De Franceschi, P., D'hallewin, G., & Bacchetta, G. (2018). Seed morphometry is suitable for apple-germplasm diversity-analyses. *Computers and Electronics in Agriculture*, *151*, 118–125. doi:10.1016/j.compag.2018.06.002

Sharma, A., Jain, A., Gupta, P., & Chowdary, V. (2020). Machine learning applications for precision agriculture: A comprehensive review. *IEEE Access : Practical Innovations, Open Solutions*, *9*, 4843–4873. doi:10.1109/ACCESS.2020.3048415

Sharma, H., Haque, A., & Jaffery, Z. A. (2019). Smart agriculture monitoring using energy harvesting Internet of Things (EH-IoT). *World Scientific News*, (121), 17–21.

Shen, W., Hu, H., Dai, B., Wei, X., Sun, J., Jiang, L., & Sun, Y. (2020). Individual identification of dairy cows based on convolutional neural networks. *Multimedia Tools and Applications*, *79*(21-22), 14711–14724. doi:10.100711042-019-7344-7

Subeesh, A., Bhole, S., Singh, K., Chandel, N. S., Rajwade, Y. A., Rao, K. V. R., Kumar, S. P., & Jat, D. (2022). Deep convolutional neural network models for weed detection in polyhouse grown bell peppers. *Artificial Intelligence in Agriculture, 6*, 47–54. doi:10.1016/j.aiia.2022.01.002

Suchithra, M., Asuwini, T., Charumathi, M. C., & Lal, R. N. (2018). Sensor data validation. *International Journal of Pure and Applied Mathematics*, *119*(12), 14327–14335.

Suchithra, M. S., & Pai, M. L. (2020). Improving the prediction accuracy of soil nutrient classification by optimizing extreme learning machine parameters. *Information Processing in Agriculture*, *7*(1), 72–82. doi:10.1016/j.inpa.2019.05.003

Sudarsan, B., Ji, W., Biswas, A., & Adamchuk, V. (2016). Microscope-based computer vision to characterize soil texture and soil organic matter. *Biosystems Engineering*, *152*, 41–50. doi:10.1016/j.biosystemseng.2016.06.006

Sujawat, G. S. (2021). Application of Artificial Intelligence in detection of diseases in plants: A Survey. *Turkish Journal of Computer and Mathematics Education*, *12*(3), 3301–3305.

Taheri-Garavand, A., Nasiri, A., Fanourakis, D., Fatahi, S., Omid, M., & Nikoloudakis, N. (2021). Automated in situ seed variety identification via deep learning: A case study in chickpea. *Plants*, *10*(7), 1406. doi:10.3390/plants10071406 PMID:34371609

Tang, J., Wang, D., Zhang, Z., He, L., Xin, J., & Xu, Y. (2017). Weed identification based on K-means feature learning combined with convolutional neural network. *Computers and Electronics in Agriculture*, *135*, 63–70. doi:10.1016/j.compag.2017.01.001

Vishwakarma, D. K., Pandey, K., Kaur, A., Kushwaha, N. L., Kumar, R., Ali, R., Elbeltagi, A., & Kuriqi, A. (2022). Methods to estimate evapotranspiration in humid and subtropical climate conditions. *Agricultural Water Management*, *261*, 107378. doi:10.1016/j.agwat.2021.107378

Weary, D. M., Huzzey, J. M., & Von Keyserlingk, M. A. G. (2009). Board-invited review: Using behavior to predict and identify ill health in animals. *Journal of Animal Science*, *87*(2), 770–777. doi:10.2527/jas.2008-1297 PMID:18952731

Xiao, J., Liu, G., Wang, K., & Si, Y. (2022). Cow identification in free-stall barns based on an improved Mask R-CNN and an SVM. *Computers and Electronics in Agriculture*, *194*, 106738. doi:10.1016/j.compag.2022.106738

Yang, Q., Xiao, D., & Lin, S. (2018). Feeding behavior recognition for group-housed pigs with the Faster R-CNN. *Computers and Electronics in Agriculture, 155*, 453–460. doi:10.1016/j.compag.2018.11.002

Yang, S., Logan, J., & Coffey, D. L. (1995). Mathematical formulae for calculating the base temperature for growing degree days. *Agricultural and Forest Meteorology, 74*(1-2), 61–74. doi:10.1016/0168-1923(94)02185-M

Zhang, C., Yue, P., Di, L., & Wu, Z. (2018). Automatic identification of center pivot irrigation systems from landsat images using convolutional neural networks. *Agriculture, 8*(10), 147. doi:10.3390/agriculture8100147

Zhao, G., Quan, L., Li, H., Feng, H., Li, S., Zhang, S., & Liu, R. (2021). Real-time recognition system of soybean seed full-surface defects based on deep learning. *Computers and Electronics in Agriculture, 187*, 106230. doi:10.1016/j.compag.2021.106230

Zhu, S., Zhou, L., Gao, P., Bao, Y., He, Y., & Feng, L. (2019). Near-infrared hyperspectral imaging combined with deep learning to identify cotton seed varieties. *Molecules (Basel, Switzerland), 24*(18), 3268. doi:10.3390/molecules24183268 PMID:31500333

Chapter 4
Plant Disease Classification in Segmented Images Using Computer Vision

Rajashri Roy Choudhury
Brainware University, India

Piyal Roy
Brainware University, India

Shivnath Ghosh
Brainware University, India

ABSTRACT

Agriculture productivity has a significant impact on the lives of people and economies because of the growing human population. In agriculture, plant diseases are a big problem since they result in severe crop losses and financial hardship for farmers. Traditional disease detection and categorization methods take a long time and are subjective, so automated and effective methods are required. Computer vision techniques have recently shown promise as tools for classifying plant diseases. To provide a precise and dependable system for disease detection and management, this article gives a thorough study on computer vision approaches for plant disease categorization. The research uses a variety of approaches, such as feature extraction, image pre-processing, and machine learning algorithms. Benchmark datasets are used for comparative study and performance evaluation of various methods. The outcomes show how effective computer vision techniques are at precisely diagnosing and categorising plant diseases.

DOI: 10.4018/978-1-6684-9975-7.ch004

1. INTRODUCTION

The goal of this research article is to establish a trustworthy system for the precise diagnosis and classification of diseases. It does this by investigating computer vision approaches for plant disease classification. Automating disease diagnosis enables prompt responses, which improves resource allocation and reduces environmental impact. In order to assess their effectiveness and applicability in agriculture, the article examines image preprocessing, feature extraction, and machine learning techniques. To overcome the drawbacks of conventional techniques, computer vision provides non-destructive, quick, and accurate illness identification. It can help with early warning systems, disease surveillance, and effective monitoring of expansive crop fields. Discussions of the methodologies used, benchmark datasets, and comparative analysis are made, assessing strengths and weaknesses. The study concludes an impact on crop management as well as potential improvements in the classification of plant diseases based on computer vision. This study advances our understanding of the subject by demonstrating how computer vision has the power to transform the way that diseases are managed, increase agricultural productivity, and guarantee food security around the world.

1.1 Background and Significance

Plant diseases are a danger to global food security and agricultural output. Traditional disease diagnosis techniques take a lot of time and are subjective. The method of identifying and categorising plant diseases has, however, been substantially enhanced by recent developments in deep learning and machine learning. The performance of convolutional neural networks (CNNs) in image recognition tasks, such as plant pathology, has been exemplary [1]. The accurate classification of plant diseases using annotated plant images is made possible by deep learning models' ability to automatically learn complicated properties from vast datasets [2]. Deep learning algorithms have several significant advantages when used to classify plant diseases. The ability to recognise infections early enables timely control actions before they spread and cause significant harm. Second, by incorporating deep learning models into intelligent agricultural systems, it is possible to monitor plant health in real-time, giving vital information for focused interventions and efficient resource use [4]. Thirdly, by enabling farmers to select suitable treatments and disease-resistant crop varieties, precise disease categorization promotes efficient disease management and yield optimisation. This lowers the need for chemicals, lessens the impact on the environment, and increases agricultural yields [3]. Fourthly, the incorporation of deep learning-based disease classification models into decision support systems might help plant pathologists and farmers make well-informed choices regarding disease control measures. Finally, accurate and efficient disease categorization models contribute to global food security by minimizing the effects of illnesses and raising agricultural production [5].

1.2 Overview of Computer Vision Techniques

Plant disease identification and classification have been made substantially easier because to computer vision techniques as deep learning, feature extraction, image segmentation, transfer learning, ensemble approaches, dataset augmentation, and interaction with IoT and sensors. Convolutional neural networks (CNNs), a type of deep learning-based algorithm, have demonstrated astounding accuracy in the identification of several plant diseases. From plant photos, disease-related patterns can be extracted and selected using features [8] [6] [3]. Techniques for image segmentation make it easier to separate and examine areas of interest related to disease. Even when there is a lack of labelled data, transfer learning uses previously trained models to learn disease-specific properties. The performance of an ensemble technique is improved by combining different classifiers or models. Training datasets are artificially increased by dataset augmentation to improve generalisation and avoid overfitting. Real-time monitoring and prompt notifications for managing diseases are made possible by integration with IoT and sensors. These computer vision approaches improve agricultural productivity and food security globally by providing efficient and effective tools for automated plant disease identification.

2. PLANT DISEASE OVERVIEW AND IMPORTANCE

Crop productivity and food security are seriously threatened by plant diseases. For effective illness management, early detection and precise diagnosis are essential. The classification and detection of diseases can be done automatically and effectively using computer vision techniques. These methods can recognise the visual signs and patterns connected to illnesses by examining digital photographs of plant parts [5]. This makes it possible to check the health of plants quickly and non-destructively, enabling prompt disease control and action. Farmers are given the information they need to make wise decisions, use resources efficiently, and promote sustainable practises thanks to computer vision. We can lessen the effect of plant diseases on the world's food output and provide a more reliable food supply by fusing technology and agriculture [7].

Figure 1. Samples of leaf images showing different types of diseases

2.1 Major Types of Plant Disease

The major types of plant diseases[1] are discussed in Table-1, Table-2, and Table-3.

2.2 Impact of Plant Disease on Agriculture

Food security, the fight against poverty, and public health are seriously threatened by the rise in plant diseases on a global scale. The spread and severity of plant diseases are influenced by several factors, including climate change, global food trade networks, and the creation of novel pathogen lineages. One such is potato blight, which led to the disastrous Irish Famine and still affects potato growers globally and necessitates major investments in fungicides [10] [11]. Plant pests have an influence

Table 1. Bacterial plant disease

Disease name	Plant affected	Causal Organism	Symptoms
Granville wilt	Tobacco, tomato, potato, eggplant, pepper, and other plants	Pseudomonas solanacearum	Stunting Yellowing Wilting of parts above ground Roots decay and become black or brown
Fire blight	Apple and pear	Erwinia amylovora	Yellowish green spots on leaves
Blight of beans	Beans (common blight)	Xanthomonas campestris	Yellowish green spots on leaves
Soft rot	Many fleshy tissue fruits such as cabbage, carrot, celery, onion	Erwinia carotovora	Soft decay of fleshy tissues that become mushy and soft
Crown gall	More than 100 genera of woody and herbaceous plants	Agrobacterium tumefaciens	Stems and roots becoming wrinkled and turning brown to black
Aster yellows	Many vegetables, ornamentals, and weeds	Mycoplasma-like organism (MLO)	Chlorosis Dwarfing malformations
Citrus stubborn disease	Citrus and stone fruits and vegetables	Spiroplasma citri (MLO)	Chlorosis Yellowing of leaves shortened internodes Wilting

Table 2. Viral or viroidal plant disease

Disease Name	Plant Affected	Causal Organism	Symptoms
Tobacco mosaic	Tobacco, tomato, and hundreds of other vegetables and weeds	Tobacco mosaic virus (TMV)	The mottled appearance of leaves (mosaic pattern) Dwarfing
Cucumber mosaic	Cucumber, bean, tobacco, and other plants (wide range of hosts)	Cucumber mosaic virus (CMV)	Similar to those of TMV infections
Barley yellow dwarf	Barley, oats, rye, wheat; also pasture grasses and weeds	Barley yellow dwarf virus (BYDV)	Yellowing and dwarfing of leaves Stunting of plants
Tomato spotted wilt	Tomato, pepper, pineapple, peanut, and many other plants	Tomato spotted wilt virus (TSWV)	Leaves show concentric, necrotic ringS Necrotic region yellow, then turning red-brown
Prunus necrotic ring spot	Stone fruits—e.g., cherry, almond, peach, apricot, plum, and others	Prunus necrotic ringspot virus (PNRV)	Delayed foliation Leaves on infected branches show light green spots and dark rings
Potato spindle tuber	Potato and tomato	Potato spindle tuber viroid (PSTV)	Stunted growth Tubers are spindle-shaped Smaller than healthy tubers
Citrus exocortis	Orange, lemon, lime, and other citrus plants	Citrus exocortis viroid (CEV)	Infected trees show vertical splits in the bark Thin strips of partially loosened bark.

Table 3. Fungal plant disease

Disease name	Plant affected	Causal Organism	Symptoms
Late blight of potato	Potato	Phytophthora infestans	Water-soaked dark green to black or purplish lesions with pale green margins on lower leaves White mildew at the edge of lesions
Chestnut bligh	Chestnut tree	Endothia parasitica	Yellowish to reddish-brown patches appear on the bark Lesions spread quickly and girdle twigs or limbs, which die
Dutch elm disease	Elm tree	Ceratocystis ulmi	Leaves wilt, turn dull green to yellow or brown and drop off Branches die
Black stem rust of wheat	Wheat and many types of grass	Puccinia graminis	On wheat, rust-coloured pustules with spores, chlorosis of surrounding tissue, followed by the development of black teliospores On barberry, chlorosis and hypertrophy of infected tissue, orange spore masses
White-pine blister rust	White Pine tree	Cronartium ribicola	Small, discoloured, spindle-shaped cankers surrounded by a narrow band of yellow-orange bark Blisters exude secretion followed by bright orange pustules
Corn smut	Corn	Ustilago maydis	Minute galls form on young corn seedlings On older plants, large galls are produced on the silk of ears and on tassels, leaves, and stalks
Citrus exocortis	Orange, lemon, lime, and other citrus plants	Citrus exocortis viroid (CEV)	Infected trees show vertical splits in the bark Thin strips of partially loosened bark
Loose smut	Barley, oats, wheat	Ustilago nuda	Infected heads are covered with masses of olive-green spores
Downy mildew	Many types of plants: grapes, grasses, vegetables, and others	many species of the family Peronosporaceae	Yellow irregular spots appear on the upper leaf surface Downy fungus growth appears on the underside
Black spot of rose	Rose	Diplocarpon rosae	Large circular black lesions on leaves Leaves turn yellow and fall off
Anthracnose of grape	Grapes	Elsinoe ampelina	Same as the black spot of rose
Nectria canker	Apple and pear and many hardwood forest trees	Nectria galligena	Initially, small circular brown areas enlarge and become depressed with raised edges Callus tissue produced around the canker
Black knot of plum and cherry	Plum and Cherry	Plowrightia morbosum	Small black knotty swellings on twigs and branches
Soft rot	Flowers, fruits, and vegetables with fleshy organs	Rhizopus species	Tissues become soft with a water-soaked appearance that often spreads rapidly, followed by the development of fuzzy grey mycelium and black spores
Fusarium wilt of tomato	Tomatoes	Fusarium oxysporum	Leaves are bent down, growth is stunted Dark streaks appear in vascular tissue

that goes beyond potatoes; it is estimated that they result in an annual loss of crop of 10–16%, or $220 billion USD. With potential losses ranging from 8% to 41%, major crops like wheat, rice, maize, potatoes, and soybeans are particularly susceptible. Considering the need to boost food production to support a growing population, this loss of food to pests and diseases is concerning. In order to guarantee a consistent food supply and enhance global health, plant diseases must be addressed.

2.3 Need for Automated Disease Classification

Deep learning algorithms and recent advancements in smartphone technology have made it possible to classify diseases automatically. In this study, photos from a smartphone are transferred to a personal computer wirelessly over a local area network (LAN) in order to classify ten distinct illnesses using transfer learning in four important but understudied agricultural crops. The related findings of six pre-trained CNNs, including AlexNet, Visual Geometry Group 16 (VGG16), VGG19, GoogLeNet, ResNet101, and DenseNet201, have been examined [12]. The need of creating autonomous plant disease diagnosis models based on discernible symptoms on leaves has recently been shown by developments in computer vision and deep learning (DL) models. Considering this, this work suggests an automated method based on an optimum mobile network-based convolutional neural network (OMNCNN) for identifying and categorising plant leaf diseases [13]. The proposed OMNCNN model uses preprocessing, segmentation, feature extraction, and classification as its primary operational steps. To pinpoint the areas of the leaf image that are impacted, bilateral filtering (BF)-based preprocessing and Kapur's thresholding-based image segmentation are used. In order to improve the accuracy of detecting plant diseases, the MobileNet model is also used as a feature extraction tool. Here, the hyperparameters are optimised using the Emperor Penguin Optimizer (EPO) algorithm [14] [13]. An innovative concept for using the images to train neural networks is the Vision Transformer (ViT)22. On certain classification datasets, such as ImageNet23, CIFAR-10024, and Oxford-IIIT Pets25, they demonstrated that this structure can achieve higher accuracy than ResNet1523 and exceed the Noisy Students (EfficientNet-L2)26 [14]. Bert27, who inspired ViT, said that "all you need is attention". There are 28 studies that use the attention mechanism for applications in natural language processing (NLP) and introduce the transformer idea. The 1D input array's sequence is received by the transformer structure. In order to handle 2D images, 2D patches are first taken from them, and after that, they are moulded to generate 1D arrays, making them suitable for ViT structures [14].

3. DATASETS AND BENCHMARKING

Benchmarking and data sets are crucial for assessing how well computer vision techniques classify plant diseases. Algorithm training and testing are made possible by specialised datasets that include annotated photos of both healthy and sick plants. The construction and assessment of machine learning models are made possible by these datasets, which give ground truth labels for the existence of diseases. Utilising standardised evaluation parameters like accuracy, precision, recall, and F1 score, benchmarking entails evaluating algorithms. Researchers can assess suggested methods' efficacy and efficiency by contrasting them with current methods on various measures. Researchers' creation of benchmark datasets encourages progress in the subject and encourages fair comparisons [17]. The development of computer vision systems for classifying plant diseases is aided by open access to a variety of datasets and standardised benchmarking procedures. They promote collaboration and stimulate creativity by making it possible to identify the advantages and disadvantages of various strategies. Researchers can create models for plant disease detection and management that are more precise and dependable by using strong datasets and benchmarking processes [17].

3.1 Available Plant Disease Datasets

To the best of our knowledge, the only publicly available dataset for plant disease identification is the PlantVillage dataset (PVD). The data set curators developed an automated approach for disease diagnosis that has a 99.35% accuracy rate using GoogleNet and AlexNet [15].

However, because the photos in the PlantVillage dataset were captured in lab settings rather than in actual agriculture areas, it is likely that they will perform poorly in the real world. In contrast, we collect actual images of healthy and ill plants to produce a dataset that is open to the public [15].

Images captured in precise conditions can be found in the PlantVillage dataset. Because actual plant photos may have several leaves with various backdrop circumstances and varied illumination conditions (as seen in Figure 1), this dataset restricts the effectiveness of disease detection. Considering this, we now describe the curated dataset and go over the curation methods [16].

In Figure 3 different plant disease datasets are shown.

3.2 Benchmarking Protocols and Challenges

The evaluation of algorithms, models, or systems for correctly identifying and categorising plant illnesses is a component of benchmarking protocols and problems

Figure 2. Leaf classes [15]

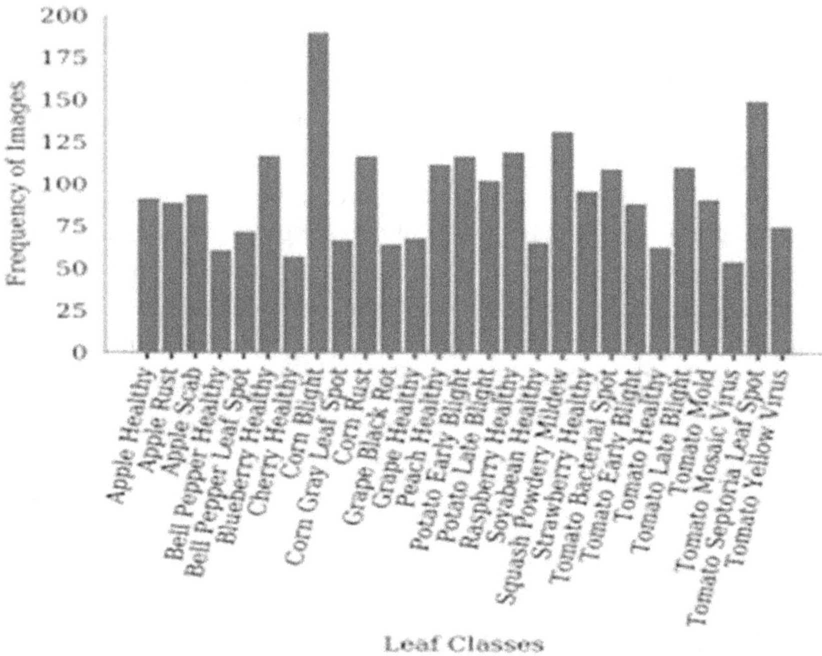

in plant disease categorization. The choice of the dataset, ensuring that different species and imaging situations are represented, and including target diseases are important considerations. Generalisation performance on omitted data is evaluated using a train-test split. Cross-validation evaluates stability and resilience across several data subsets. Performance indicators that measure classification system effectiveness include mAP, accuracy, precision, recall, F1-score, and AUC-ROC. Finally, contrasting results with benchmark or cutting-edge models reveals improvements and system utility [17].

4. STATE OF THE ART APPROACHES AND METHODS

Modern methods have greatly aided in the classification of plant diseases. CNNs have shown exceptional performance in precisely diagnosing plant illnesses. Automated illness identification is made possible by these models, which extract discriminative features from unprocessed visual data. High classification accuracy can also be

Figure 3. Different plant disease datasets [17]

Table 4. Transfer learning doubled the accuracy after finetuning on uncropped plantdoc dataset [15]

PreTrained Weights	Training Set (Set%)	Test Set (Set%)	Accuracy	F1-Score
Image Net	PlantDoc(80)	PlantDoc(20)	13.74	0.12
Image Net	PVD	PlantDoc(100)	15.08	0.15
ImageNet+PVD	PlantDoc(80)	PlantDoc(20)	29.73	0.28

attained via transfer learning techniques, which include fine-tuning pre-trained CNN models using datasets related to plant diseases [2] [7] [18].

Another strategy is to combine machine learning algorithms with computer vision techniques. While feature extraction techniques like GLCM and texture analysis capture useful information from these segmented regions, image segmentation techniques like K-Means clustering and spot detection algorithms assist in isolating disease-affected regions. The collected features are subsequently classified using machine learning algorithms, such as Support Vector Machines (SVM), Random Forests, and ensemble methods, to detect certain diseases [18][19].

4.1 Traditional Computer Vision Techniques

Traditional computer vision techniques have been the foundation of image analysis and understanding for decades. These techniques rely on extracting features from images and applying algorithms for tasks like object detection and recognition. They often involve steps such as edge detection, image segmentation, and feature extraction.

4.1.1 Image Acquisition

Image acquisition is the process where Images of plant leaves that need to be examined for illnesses are sent to the software during this stage of the procedure. Normally coloured, the images could include different details. However, the photos are changed to grayscale to make it simpler to diagnose and analyse the disorders. Images that are grayscale are black and white, meaning that they lack colour and only have shades of grey ranging from black to white. The classification process is made simpler by converting the photographs to grayscale because the programme can concentrate exclusively on changes in shades of grey rather than having to cope with different colours. By simplifying the process, it becomes simpler to spot patterns and anomalies in the leaf structure that may point to the existence of a disease. The software accesses the plant snapshot and imports the image into its database for additional analysis. The photos are stored in JPEG format, which is a popular file format for digital image storage. Smaller file sizes are possible using JPEG compression while maintaining a respectable degree of image quality. The software maintains uniformity in image analysis by using standard resolutions, making it simpler to compare and categorise various photographs of plant leaves [20] [21]

4.1.2 Image Pre-Processing

Image pre-processing is the step involves manipulating the leaf images to improve their quality and visibility. Techniques like resizing the images to a standard size and enhancing their contrast are commonly employed. These adjustments help in standardizing the images and enhancing the features of interest. Image preprocessing is a set of techniques, that are used to modify or enhance images before they are subjected to further analysis or processing. One of the common preprocessing techniques is the unsharp filter, which is used to sharpen an image by enhancing

Figure 4. General architecture of the plant disease classification system

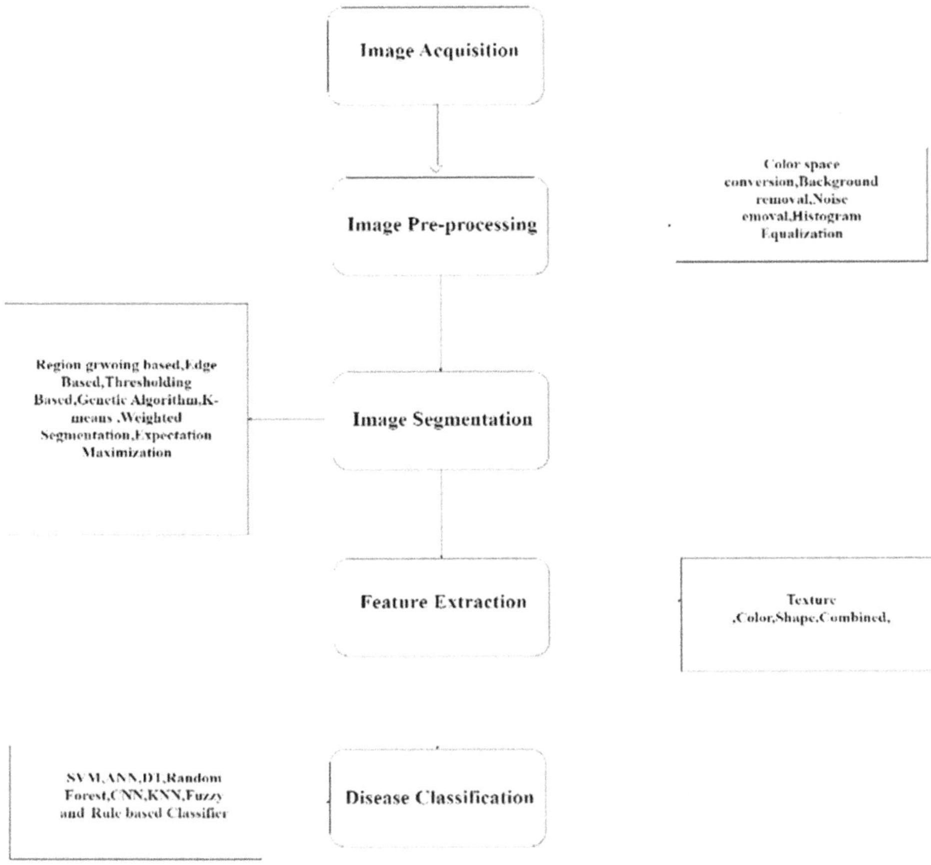

its edges and high-frequency components. The name "unsharp" filter refers to the process of removing the smoothed or blurred version of the original image [20].

4.1.3 Segmentation

Segmentation is the technique of isolating the leaf regions from the foreground or other objects in an image for the purpose of classifying plant diseases. To distinguish the leaf from the rest of the image, K-Means clustering, which groups comparable pixels based on colour or intensity values, is frequently utilised. The quality of the leaf photos is enhanced in the Figure 5 model by pre-processing methods such histogram equalisation and image enhancement [8][2][3][22][23]. Several processes are involved in the entire illness classification model, including locating and separating the leaf region from the background, extracting features from the leaf region, and using these features to classify the diseases [22].

Deep learning has recently improved disease localization by segmenting data using the U-Net architecture. A well-known CNN architecture created especially for image segmentation tasks is called U-Net. It has a U-shaped structure and is made up of an encoding path and a corresponding decoding path [23]. While the decoding path recovers spatial information and produces the segmented output, the encoding path collects features and captures context from the input image [23].

The performance of various U-Net implementations and a modified version is investigated in the study. The design consists of an expanding path that mixes up-sampled features with the features from the contracting path for segmentation and a contracting path for down-sampling through convolutions and pooling operations. There are numerous convolutional layers in the network [23]. This study's updated U-Net model performs better by adding three convolutional layers in place of two in the decoding blocks. A smaller number of photos is processed during training, then a larger set is processed during the upsampling stage. SoftMax is a technique used to categorise background and object pixels in the model's final layer [23].

4.1.4 Feature Extraction

Feature extraction means once the leaf regions are segmented, features need to be extracted to represent the unique characteristics of the diseased areas. One common method is using the Gray-Level Co-occurrence Matrix (GLCM). GLCM calculates the texture properties of the segmented regions, such as contrast, energy, and entropy, which can be indicative of the presence of diseases.

Figure 5. Flow chart explaining image localization and classification process [22]

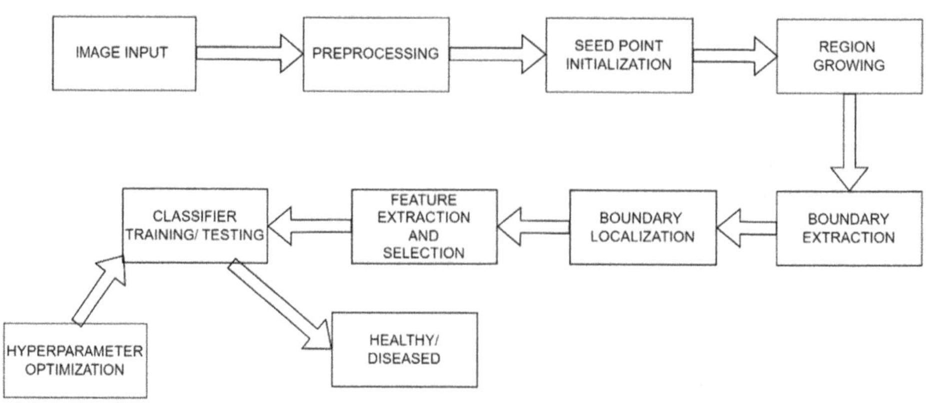

Figure 6. U-net architecture

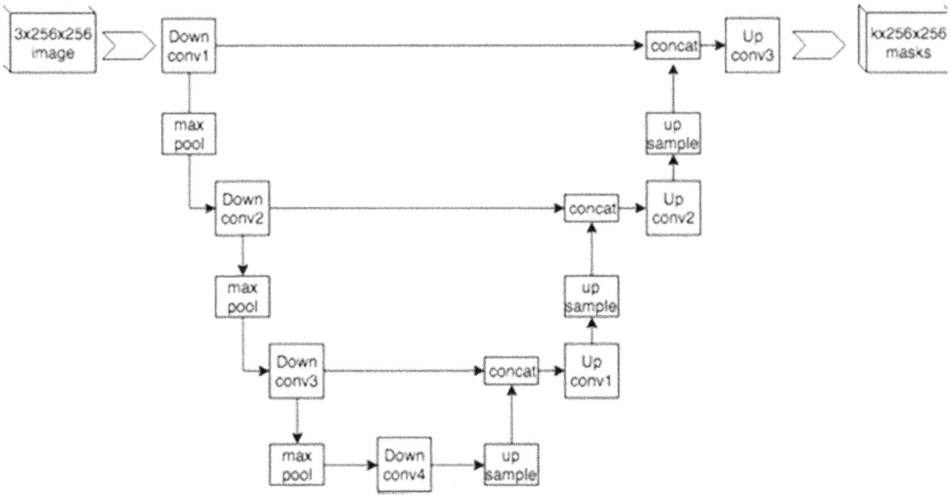

Extraction of shape features: In this study, shape features like eccentricity, extent, solidity, and minor axis length were utilised. The sick area of the leaf under consideration is extracted using these features [20].

Extraction of texture features: The texture features used in the work include contrast, correlation, and energy. These characteristics are used to identify the sick area of the leaf under consideration. The variation of pixels and their neighbouring pixels will then be determined [20].

Colour feature extraction: The colour feature extraction has a special manner of displaying image representation when it comes to translation, scaling, and rotation. Mean, skewness, and kurtosis are the characteristics used for colour [20].

Here, RGB is converted to LAB [20].

Deep feature extraction: Utilising a pre-trained neural network to retrieve learnt characteristics from particular layers is known as feature extraction. For instance, VGG16 may extract features from the most recent completely connected layers, whereas ResNet50 creates a feature vector by flattening extracted features. In its "pool 5 drop 7x7 s1" layer, GoogleNet employs maximum pooling. Through the use of pre-trained networks' expertise, feature extraction accelerates the process of extracting information from images [20].

i. Feature Extraction Results with Resnet 50, Google Net and VGG16

The study discovered that the fully connected layer 6 (fc-6) of the VGG16 network was the most efficient for feature extraction in plant disease classification using SVM and KNN algorithms. With SVM, GoogleNet performed better while ResNet50 had a greater overall accuracy but required more time to train. For a thorough understanding of the outcomes, refer to Table-5 and Figure-7 for more specific performance indicators [24].

ii. Deep Learning Results with Resnet 50, Google Net and VGG16

By transferring knowledge from one model to another, transfer learning increases the effectiveness of learning. Knowledge can be transferred by fine-tuning a network that has already been trained. The necessity for freezing layers to avoid overfitting depends on the size of the target dataset. Performance is enhanced via data augmentation, callbacks, and the proper optimisation methods. VGG16 had the best accuracy (97.92%) out of the studied networks (GoogleNet, ResNet50), although it took longer to execute. In 12 minutes and 30 seconds, GoogleNet achieved an accuracy of 95.30 percent. In terms of execution speed and accuracy, ResNet50 came in third. The true positive rate, false positive rate, and F-score were examined performance indicators. Effective information transfer is facilitated by transfer learning, which also improves model performance [24].

iii. Performance Based on Traditional Shallow Networks

Figure 7. Graph displaying the results based in feature extraction with SVM and KNN

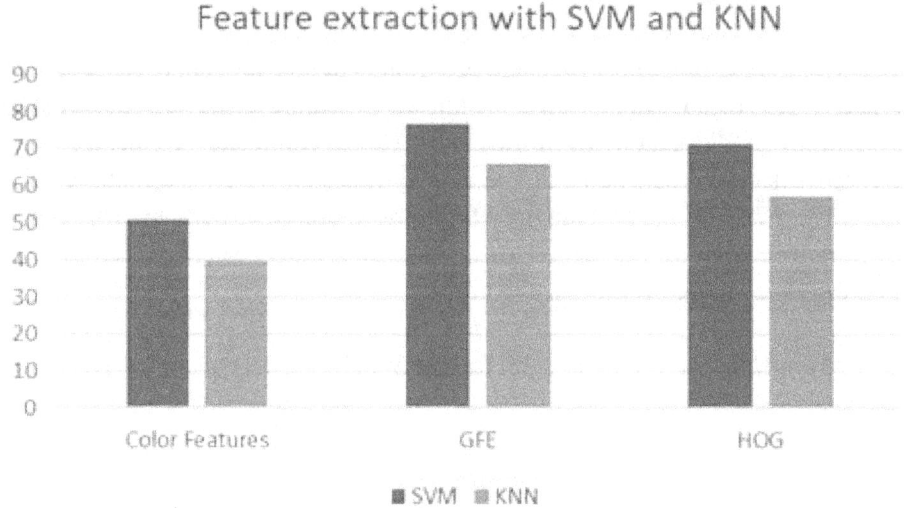

Figure 8. Results based on transfer learning

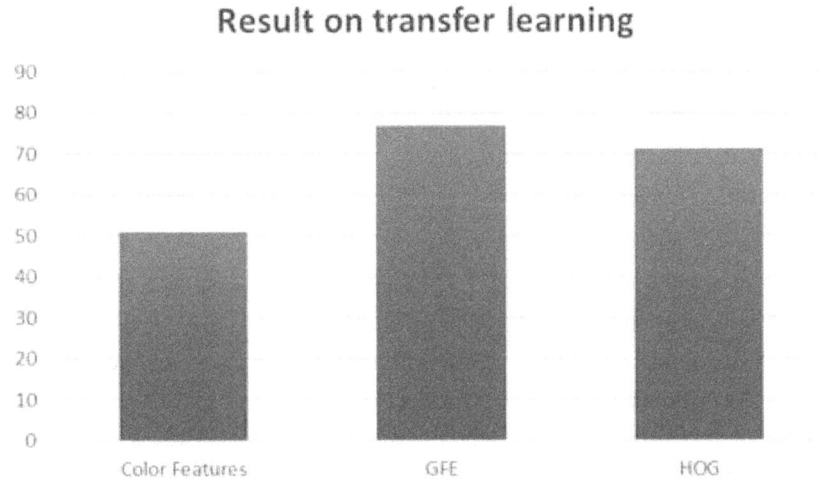

Table 6. Performance of the models in transfer learning, in accuracy and time

	VGG16	Google Net	Resnet 50
Accuracy(%)	97.82	95.3	95.38
Time	14 min 32s	12 min 30s	23 min 49s

Table 5. Performance of the models according to true positive rate, false positive rate and F score measures

Classifier	Performance Measure	VGG16	Google Net	Resnet 50
SVM	True positive rate False Positive Rate F score	94.01 99.20 94.18	94.30 98.20 94.45	97.50 99.43 97.13
KNN	True positive rate False Positive Rate F score	87.49 99.45 86.36	87.80 98.60 87.90	87.90 99.02 88.04

We used a number of feature extraction techniques, including colour features (CF), gamma based feature extraction (GFE), histogram of oriented gradients (HOG), and local binary pattern (LBP), to identify plant illnesses. These methods assisted in gaining useful information from the photos. In order to assess the accuracy of the collected features, Support Vector Machine (SVM) and K-Nearest Neighbours (KNN) classifiers were used. According to Table 8's findings, LBP and the SVM classifier together produced the maximum accuracy of 80.6%. GFE and HOG both obtained

Table 7. Measuring true positive rate, the false positive rate and f-score using transfer learning

Performance	VGG16	Google net	Resnet 50
True positive rate	97.20	93.29	94.91
False positive rate	99.60	96.75	98.80
F-score	96.42	93.54	93.18

accuracy with SVM of 76.9% and 71.28%, respectively. With 51.03% accuracy, colour characteristics performed substantially less accurately. These findings are illustrated visually in Table 8 and Figure 9. In simple term, we employed a variety of approaches to extract features from photos of plant diseases. The most accurate was LBP, followed by GFE and HOG. The accuracy of colour features was lower than that of the other methods. These findings aid in our understanding of the efficacy of various feature extraction techniques for identifying plant diseases [24].

4.1.5 Classification

To identify the type of leaf disease present, classification entails giving a label or category to the retrieved data. For this purpose, a variety of classifiers are employed,

Figure 9. Results based on traditional shallow methods

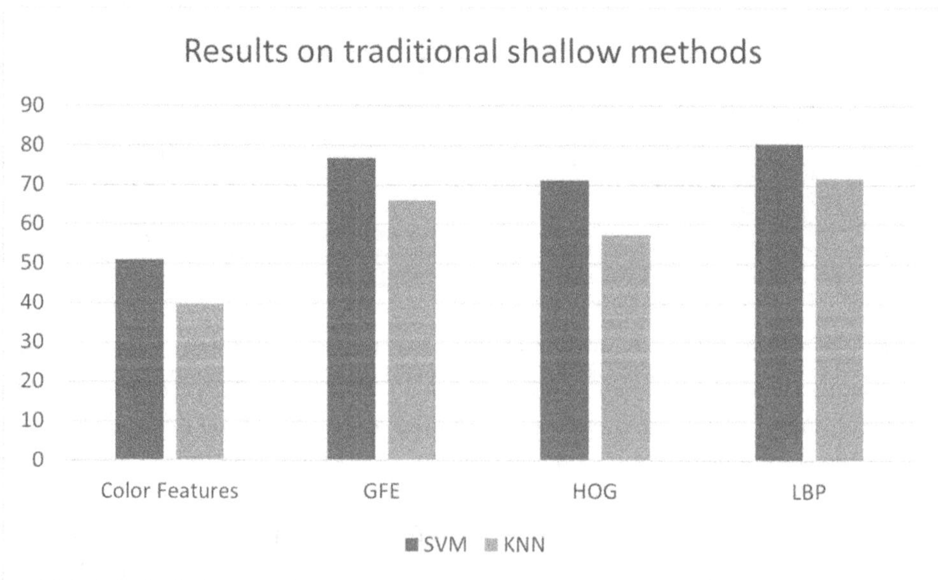

Table 8. The accuracy of traditional methods

Classifier Methods	Color Features	GFE	HOG	LBP
SVM	51.03	76.90	71.28	80.60
KNN	39.70	66.12	57.30	71.80

including Support Vector Machines (SVM), K-Nearest Neighbours (KNN), and Artificial Neural Networks (ANN). For precise prediction, these classifiers identify patterns and correlations in the retrieved features by training on labelled data [8].

Data must be divided into two genres for classification—training sets and testing sets. For each instance or piece of data in the training set, there is one target value and several characteristics. Finding the separation hyperplane that will classify these points into two distinct groups—the positive groups as "+1" and the negative groups as "-1"—is the key task at hand [20].

4.2 Convolutional Neural Network

CNNs are a type of deep neural network created especially for the analysis of visual input, such as images and movies. In a variety of applications, it has transformed computer vision tasks and produced ground-breaking outcomes. Convolution, pooling, and fully connected layers make up a CNN, which is modelled after the visual cortex in the human brain and applies operations to the input data. In order to perform tasks like predicting and classification, CNNs use the data that is provided as input to their structures. CNN uses activation functions (AFs) to compute feature maps. Mathematically the function is defined as [25].

$$y_j^l = f\left(z_j^l\right)$$

where y_j^l is called the future map and $f(z_j^l)$ is called the AF [25].

Using 2-dimensional convolution operation, CNN stores a given dataset. The length of output O, mathematically calculated as below [25]

$$O = \frac{\left(W - F + 2P\right)}{\left(S + 1\right)}$$

where W stands for input length, F for filter size, S for stride, and P for padding. Generally, as figure 10 depicts, a typical neural network has four main layers: convolutional layer, pooling layer, AF, and fully connected layer.

4.2.1 Convolutional Layer

The matrix technique is used in this layer to extract feature maps from the source image. In Figure 11, this mathematical procedure is shown. The filter is first gradually moved, beginning in the upper left corner of the image. The values of the filter (kernel) are multiplied by the values of the image at each step, and the result is then added up. The supplied image is used to generate a smaller new matrix. A convolution is described mathematically as the result of two functions, f and g, which are the subjects of the algebra of Schwartz functions in Rn. A convolution of two functions f and g over infinite range $[-\infty, \infty]$ is given by [25].

$$(f * g)(t) \overset{def}{=} \int_{-\infty}^{\infty} f(\tau) g(t-\tau)\, d\tau$$

where (f∗g) (t) denotes the convolution of functions f and g.

4.2.2 Pooling Layer

After the convolutional layer operation is finished, the pooling layer operation starts. The process of pooling, often referred to as down sampling, is fascinating. It develops the dominant response in this small region by using similar information from the receptive fields of the neighbourhood [69, 73]. The operation uses a variety of functions to achieve this, including max pooling, average pooling, and sum pooling.

Figure 10. Sample CNN architecture [25]

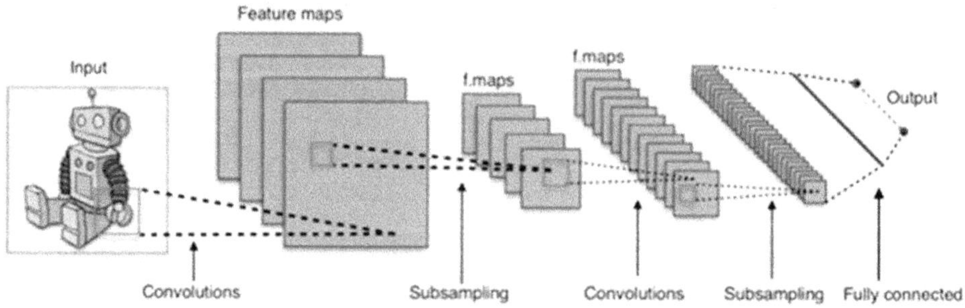

Figure 11. Convolution operation

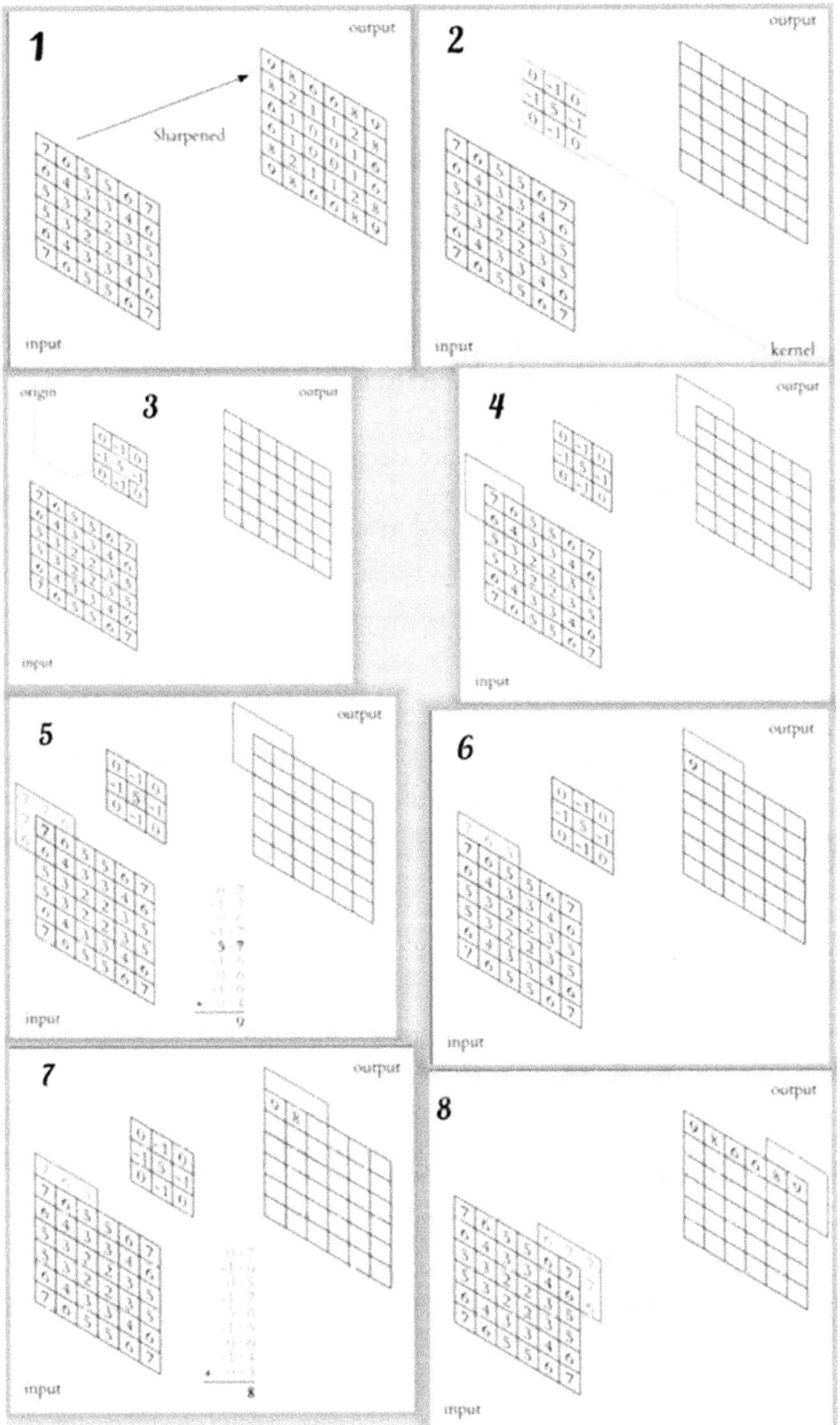

For instance, the max pooling operates by choosing the largest element from the input matrix for the filter. The operation is illustrated in Figure 12 [25]

4.2.3 Activation Function (AF)

The learning of complicated patterns and the rate of neural network convergence are influenced by the activation function (AF) selected. Several AFs are utilised, including leaky ReLU, ELU, PReLU, sigmoid, tanh, maxout, SWISH, and ReLU. The linear regression equations that emerge from neural networks are prevented by AFs. Given that they solve the vanishing gradient issue, ReLU and its variants are recommended. On numerous benchmark datasets, MISH has outperformed ReLU. The three most popular AFs are Softmax, Relu, and Sigmoid [25].

A) A) Softmax AF: AF Softmax. Deep learning algorithms typically employ it in the output layer to base choices on the weight of the input variable. The mathematical definition of Softmax AF is

$$\sigma\left(\vec{z}\right)_i = \frac{e^{z_i}}{\sum_{j=1}^{K} e^{z_j}}$$

B) B) Rectified Linear Unit (ReLU) AF: Numerous deep learning research primarily employ the ReLU AF and its variations [79]. A representation of the ReLU AF is

Figure 12. Max pooling [25]

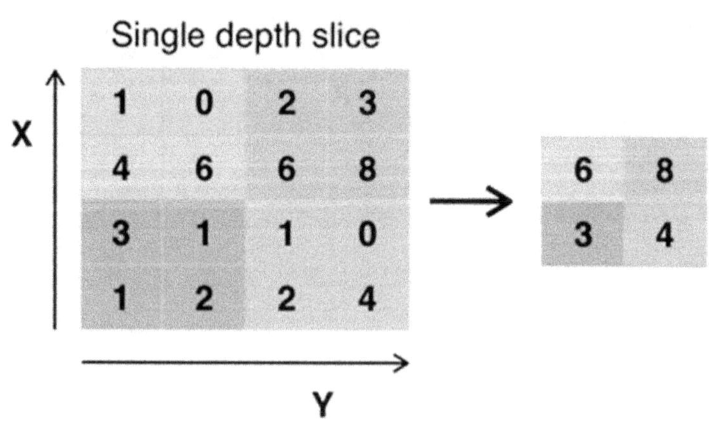

$$f\left(x\right) = \max\left(0, x\right) = \begin{cases} x \; if x \geq 0 \\ 0 \; if x < 0 \end{cases}$$

C) C) Sigmoid AF: It is one of the most prevalent AF [79] and makes decisions between 0 and 1 using a probabilistic technique. The definition of Sigmoid AF in mathematics is

$$G\left(z\right) = \frac{1}{\left(1 + e^{-z}\right)}$$

4.2.4 Fully Connected (FC) Layer

A basic feed-forward neural network serves as the FC layer. The flattened output of the flattened final pooling or convolutional layer serves as an input for the FC layer. The procedure of transforming the three-dimensional matrix output of a pooling or convolutional layer into a one-dimensional array is referred known as "flattening." Figure 13 shows the transformation of a 3D matrix into a 1D array.

The methodology seeks to classify images according to a particular viewpoint. It requires little pre-processing compared to other neural network algorithms and recognises channels that are manually created during calculations. Convolutional layers and parameter layers with constrained open fields found in a CNN's core square allow the algorithm to recognise and classify highlights [25].

A) Inception - V4: A 48-layered deep CNN network called Inception V4 expands on the ImageNet paradigm. It consists of asymmetric and symmetric blocks that use convolutions to build feature maps, average computation, maximum pooling, and average pooling (8x8). Neurons are connected by fully connected layers, and overfitting is reduced via pooling and dropouts. To calculate losses and generate activation norms, softmax and batch norm are employed [26].
B) VGG-16: VGG, also known as "visual geometry group," was developed by Oxford University for the 2014 "Image Net Large Scale Recognition of Visual Challenge" (ILSVRC). When compared to other deep neural networks, this structure has the best functionality currently available. Even though there are many different types of parameters, this structure solely concentrates on the padding convolution layer, completely connected layer, and max pool with a softmax layer as the result [26].

Figure 13. FC layer operation

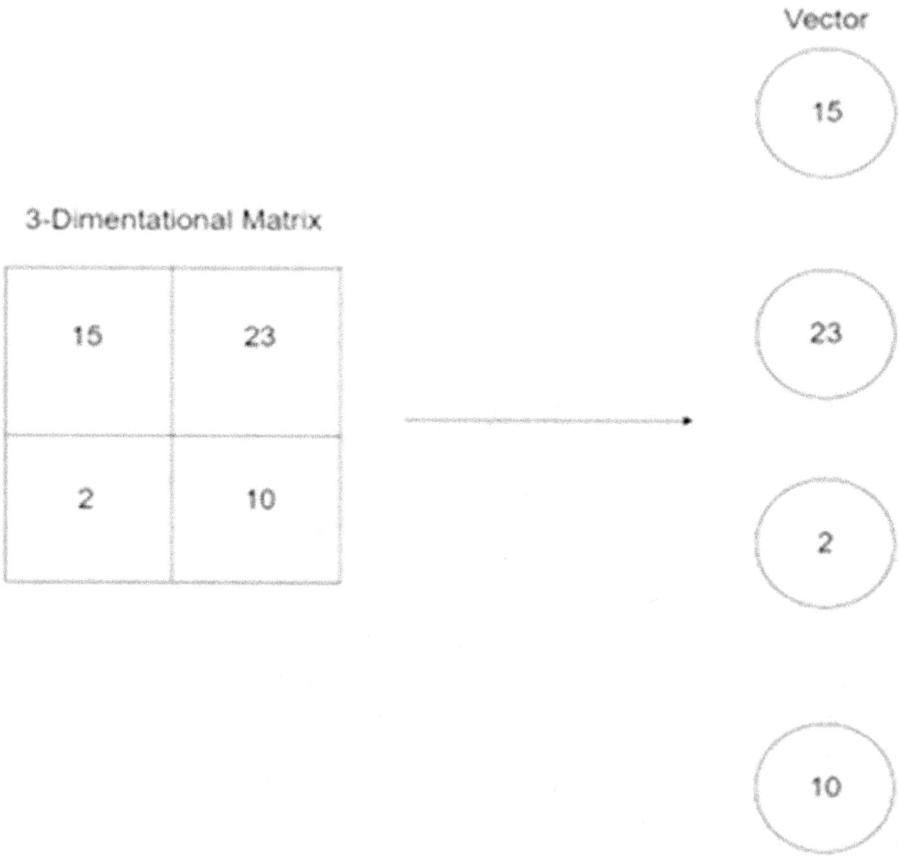

C) VGG – 19: Similar to VGG-16 and the other VGG variations, this one also has three convolution layers as an added feature, which effectively enables image recognition. The fundamental principle behind constructing deep neural networks is to use modest size and constant convolution [26].

4.3 Transfer Learning and Pre-Trained Model

A machine learning technique called transfer learning uses information gained from one task or dataset to enhance performance on a related but unrelated assignment. Transfer learning expedites model construction, improves performance, and requires less data by training a model on a source task and then optimising it on a target task. Pre-trained models from big benchmark datasets like ImageNet are frequently utilised in computer vision [27]. These models, like VGG, Inception, and MobileNet, are imported and used, which reduces the cost of computation. Using data from the ImageNet challenge, Canziani et al. (2016) present a thorough examination of pre-trained models' performance on computer vision tasks. Building on prior knowledge and standing on the shoulders of giants in deep learning are made possible through transfer learning. When adapting a pre-trained model for personal requirements, it is necessary to eliminate the original classifier initially and incorporate a new classifier that suits the specific needs. Subsequently, fine-tuning the model can be accomplished through one of three approaches.

Train the Entire Model: Using the pre-trained model's architecture, the model is trained using a vast dataset and significant computing resources in this situation. Training the model from scratch requires a large dataset.

Train Some Layers and Leave the Others Frozen: The higher layers pertain to specific aspects relevant to the problem at hand, while the lower layers capture general features that are problem-independent. In this approach, we manipulate this distinction by determining the extent to which we modify the network's weights. Frozen layers remain unchanged during training, usually to prevent overfitting when the dataset is limited and there are numerous parameters. However, if the dataset is substantial and the parameter count is low, overfitting is not a concern, and the model can be improved by training additional layers for the new task.

Freeze The Convolution Base: This instance represents an extreme application of the train/freeze trade-off. The key concept is to maintain the convolutional base in its original configuration and then feed the classifier with its results. If the dataset is tiny, computational resources are limited, or the pre-trained model already has a solution to the problem, it can be utilised as a fixed feature extraction strategy.

4.3.1 Transfer Learning Process

There are several pre-trained models available when utilizing transfer learning, including VGG, InceptionV3, and ResNet. The problem's nature and how closely the dataset resembles the dataset used for pre-training will determine which model to use. Figure 15's depiction of "The Matrix" can help with the decision-making. Based on the size of the dataset and how closely it resembles the dataset used to train the pre-trained model, this matrix categorizes computer vision difficulties. Consider utilizing Strategy 3 if the dataset is modest (less than 1000 images per class) and comparable to the pre-trained model's dataset.

Strategy 1 can be used to build a model from scratch in Quadrant 1 or the pre-trained model's architecture and weights can be used when the large dataset is distinct from the pre-trained model's dataset.

Figure 14. Fine-tuning strategies

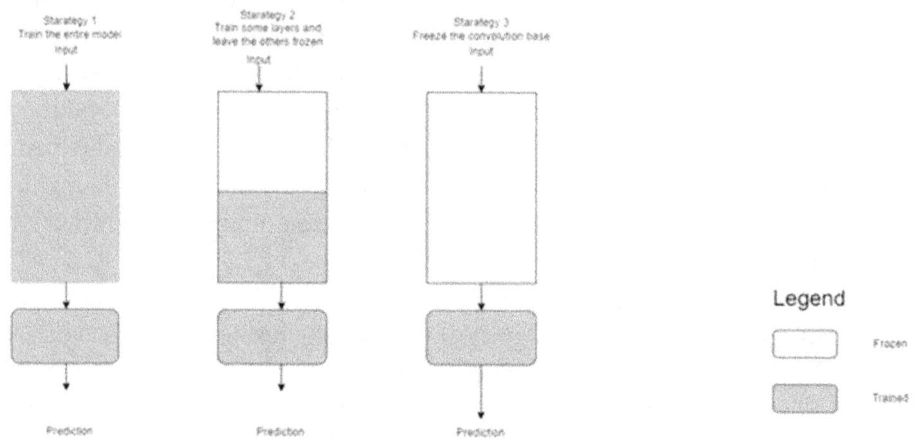

In Quadrant 2, where there is a huge dataset akin to the pre-trained model's dataset, there is more leeway. It is advised to use Strategy 2, in which the classifier and top layers of the pre-trained model are fine-tuned while utilizing the existing knowledge without extensive training.

A small dataset that is different from the pre-trained model's dataset poses a problem in Quadrant 3. The only alternative left in this situation is Strategy 2, although it can be challenging to strike the proper balance between training layers and freezing layers to prevent underfitting or overfitting.

The best option is Strategy 3 in Quadrant 4, where the small dataset is comparable to the pre-trained model's dataset. A new classifier can be trained using the generated features and the pre-trained model as a fixed feature extractor.

4.4 Ensemble and Fusion Technique

Ensemble learning is the process of classifying the outputs with a final class label using a voting mechanism. The outputs are acquired by building many classification models utilising diversity and supplying the sample as input to these models. Numerous studies have demonstrated that compared to a single classifier, the ensemble learning strategy offers a greater accuracy rate. The Bagging method's algorithm, which is a fundamental ensemble learning technique. Here, the information source required for each classifier is obtained by the Bootstrap method. The majority voting method is used at the output stage [28].

Figure 15. Size-similarity matrix [27]

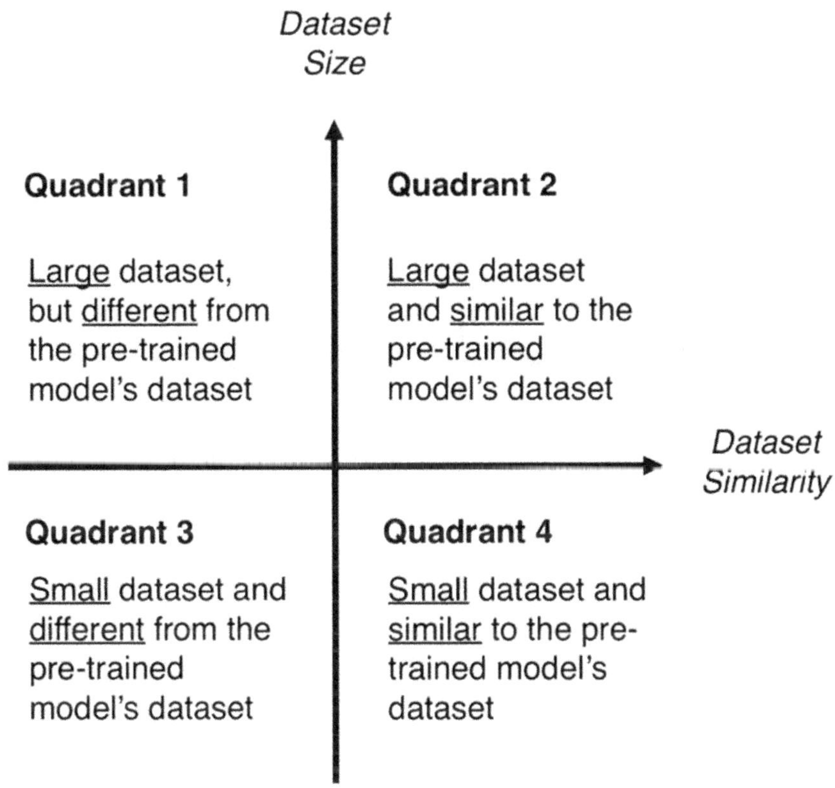

4.5 Explainable AI and Interpretability

Several strategies can be used to improve the interpretability of AI models for plant disease categorization. In order to specifically describe how different plant diseases are identified based on ocular cues in segmented images, rule-based models can be created using preset criteria. By assessing the contribution of each feature or region to the model's decision, it is possible to evaluate the importance of image features and get knowledge of the characteristics that are essential for disease categorization. With the help of visualisations like saliency maps and heatmaps, significant areas of the segmented image can be highlighted, providing clear proof of the presence of a disease. Local interpretability can be attained by creating per-instance explanations that describe the traits or regions impacting disease classification using techniques like LIME or SHAP [29] [30].

Figure 16. Decision map for fine-tuning pre-trained models [27]

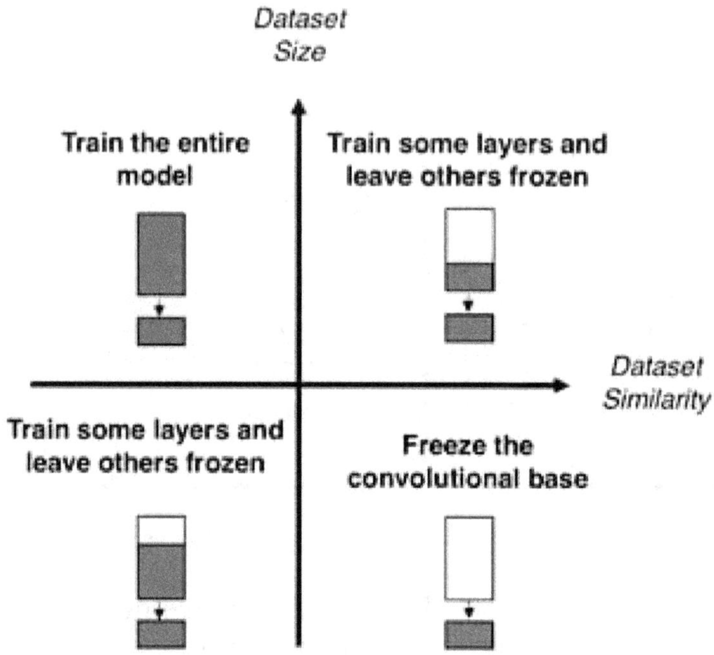

5. PERFORMANCE EVALUATION AND COMPARATIVE ANALYSIS

Researchers use performance evaluation and comparison analysis to compare and assess the efficacy of various plant disease categorization approaches. Accuracy, F1 score, recall, and precision are common evaluation metrics that rate how accurately and completely diseases are classified. By separating the dataset into subgroups for training and testing, cross-validation techniques are frequently employed to assure fair comparisons [31]. Comparative analysis compares various methods using the same benchmark dataset to determine which ones are the most precise and efficient. This analysis aids in the choice of the best methods for classifying plant diseases by enabling researchers to evaluate the benefits and drawbacks of various approaches.

5.1 Comparative Studies of Different Approaches

In this study, various deep architectures for detecting plant leaf disease are researched and evaluated. The 20640 image plant village dataset, which comprises 15 classes and 3 species—pepper, potato, and tomato—is used to carry out the investigation.

The dataset's total photos are split into three sets: 70% are used for training, 20% are used for validation, and 10% are used for testing. The experiment is conducted using MATLAB R2019b. Initial variation in the number of epochs and learning rate determines the proper value of the hyper parameter [32]. Also, by training CNN models on segmented image data, this research investigates a potential fix for this issue. When tested on independent data previously unknown to the models, even with 10 illness classifications, the S-CNN model trained using segmented images more than doubles in performance to 98.6% accuracy as compared to the F-CNN model trained using whole images. Additionally, we demonstrate that the S-CNN model greatly outperforms the F-CNN model in terms of self-classification confidence by utilising a tomato plant and a target spot disease kind as an example.

5.2 Evaluation Metrics and Performance Measures

When evaluating the efficacy and performance of plant disease classification models, evaluation metrics and performance measures are essential. These metrics offer numerical evaluations of the model's precision, robustness, and dependability. Here are some evaluation metrics and performance measurements that are frequently used in the classification of plant diseases [9] -

Accuracy: The number of cases that were successfully classified relative to the total number of examples serves as a proxy for accuracy, which assesses the overall accuracy of the classification predictions. It is characterised as

Accuracy = (TP + TN) / (TP + TN + FP + FN)

where TP = True Positive (correctly classified positive instances),

TN = True Negative (correctly classified negative instances),

FP = False Positive (incorrectly classified positive instances),

FN = False Negative (incorrectly classified negative instances)

Precision: It is the percentage of instances accurately categorised as positive out of all instances projected to be positive. When the goal is to reduce false positives, it is helpful.

Precision is equal to TP / (TP + FP)

Recall (Sensitivity or True Positive Rate): Recall quantifies the percentage of positively classified instances that are accurately classified out of all positively classified instances. When minimising false negatives is the main goal, it is helpful.

Recall = TP / (TP + FN).

F1 Score: The F1 score strikes a balance between precision and recall in a single metric. It is calculated as follows to determine the harmonic mean of recall and precision:

F1 Score is equal to 2 * (Precision * Recall) / (Precision + Recall).

Specificity: The percentage of accurately categorised negative events among all actually negative instances is known as specificity [28] (also known as the true negative rate). When the goal is to reduce false positives, it is helpful.

Specificity is TN / (TN + FP)

Area Under the Receiver Operating Characteristic Curve (AUC-ROC): By calculating the area under the ROC curve, AUC-ROC assesses the effectiveness of a classification model. At different classification thresholds, the ROC curve compares the true positive rate (sensitivity) against the false positive rate (1 - specificity). An improved classification performance is shown by a higher AUC-ROC value.

Confusion Matrix: A confusion matrix shows how the model's predictions stack up against the actual labels from the ground truth in a tabular format. It allows for a thorough comparison of the model's performance across several classes because it includes counts for true positives, true negatives, false positives, and false negatives [25].

Mean Average Precision (mAP): mAP is frequently employed in activities involving the localisation and detection of objects. After calculating the mean over all classes, it takes the average precision for each class. It offers a thorough assessment of the model's performance by considering both precision and recall at different confidence thresholds [8] [5].

Cohen's Kappa: A measure of inter-rater agreement, Cohen's Kappa considers the likelihood of agreement by chance and the agreement between the model's predictions and the ground truth labels [15].

5.3 Strengths and Limitations of Existing Methods

Deep learning techniques for classifying plant diseases are highly accurate and resilient because they can automatically learn hierarchical representations from vast volumes of data. They excel at handling intricate and varied image features, allowing for accurate disease identification and classification across many plant species. When taught with enough labelled data, deep learning models can recognise complex patterns and deliver cutting-edge performance.

Limitation is since deep learning models need a lot of labelled examples to train, the main restrictions stem from the requirement for huge annotated datasets. Such datasets can be expensive and time-consuming to acquire and annotate. Deep learning models may also have interpretability issues, making it difficult to comprehend the thinking behind their judgements. Additionally, overfitting can happen when the model fails to generalise to new data and when working with small or unbalanced data sets [34] [35].

6. APPLICATIONS AND FUTURE SCOPE

Plant pathology, precision agriculture, and crop management are just a few of the fields where deep learning in plant disease classification has found use. It makes it possible to identify diseases accurately and early, which makes it easier to manage diseases, intervene quickly, and maximise yield. Farmers, academics, and policymakers can benefit from the deep learning algorithms that have been used to classify diseases based on aerial and leaf-level images [36].

Deep learning presents intriguing opportunities for the classification of plant diseases in the future. Large-scale disease surveillance can be improved by integration with remote sensing technology like satellite imaging and drones. Even with little labelled data, transfer learning and domain adaption strategies can boost model performance. AI techniques that are explicable will make them interpretable, promoting adoption and trust. Furthermore, improvements in edge computing and low-power hardware may enable farmers to make decisions in real time by bringing these models closer to on-field deployment.

6.1 Application of Plant Disease Classification

Plant disease classification has numerous and significant real-world applications. Automated disease classification systems can be used by farmers and agricultural organisations to identify and manage illnesses, improving crop output and lowering financial losses. Models for classifying plant diseases are useful in precision farming because they help with resource allocation and focused pesticide application. Additionally, by enabling early disease identification, halting the spread of illnesses, and facilitating prompt interventions, plant disease classification supports global food security. These applications empower farmers with valuable insights, facilitate sustainable agriculture practices, and support the development of resilient and productive crop systems [38] [39].

6.2 Emerging Trends and Future Scope

Plant disease classification is poised to develop thanks to emerging trends and future directions. Comprehensive disease monitoring is made possible by the integration of multimodal data sources like hyperspectral imaging and IoT-based sensors. With little labelled data, transfer learning and deep neural networks provide efficient illness classification. AI methods that are easy to understand encourage openness and interpretability. Real-time illness diagnosis in the field is made possible through the usage of edge computing and mobile applications. Researchers and farmers that work together and share data can create disease categorization models that are reliable and relevant all over the world. Precision agriculture, sustainable farming methods, and improved global food security are all being driven by these trends. [39] [37]

6.3 Challenges and Potential Solutions

There are various difficulties in classifying plant diseases that call for creative solutions. The training and generalisation of models are hampered by small and unbalanced datasets. Transfer learning, data augmentation methods, and active data gathering initiatives are needed to address this. Deep learning models are still difficult to interpret, necessitating explainable AI techniques to provide us insights into how decisions are made. Edge computing and lightweight models are necessary to get beyond on-field deployment's hardware and computational constraints. Domain adaptation mechanisms are required for plant species diversity and robustness to environmental changes. To share data, standardise data, and validate models, researchers, farmers, and policymakers must work together. These techniques help create plant disease categorization systems for sustainable agriculture that are more precise and useful.

REFERENCES

Akbar, W., Soomro, A., Ullah, M., Haq, M. I. U., Khan, S. U., & Shah, T. A. (2023, March). Performance Evaluation of Deep Learning Models for Leaf Disease Detection: A Comparative Study. *In 2023 4th International Conference on Computing, Mathematics and Engineering Technologies (iCoMET)* (pp. 01-05). IEEE. 10.1109/iCoMET57998.2023.10099223

Arsenovic, M., Karanovic, M., Sladojevic, S., Anderla, A., & Stefanovic, D. (2019). Solving current limitations of deep learning based approaches for plant disease detection. *Symmetry, 11*(7), 939. doi:10.3390ym11070939

Arsenovic, M., Karanovic, M., Sladojevic, S., Anderla, A., & Stefanovic, D. (2019). Solving current limitations of deep learning based approaches for plant disease detection. *Symmetry, 11*(7), 939. doi:10.3390ym11070939

Ashwinkumar, S., Rajagopal, S., Manimaran, V., & Jegajothi, B. (2022). Automated plant leaf disease detection and classification using optimal MobileNet based convolutional neural networks. *Materials Today: Proceedings, 51*, 480–487. doi:10.1016/j.matpr.2021.05.584

Aswini, M., & Kavitha Bagya, S. (2022). Phytoconstituent analysis of Alpinia galanga and Zingiber officinale plant lipids–A comparative approach. *International Journal for Research Trends and Innovation*, *7*(8), 647–654.

Borhani, Y., Khoramdel, J., & Najafi, E. (2022). A deep learning based approach for automated plant disease classification using vision transformer. *Scientific Reports, 12*(1), 11554. doi:10.103841598-022-15163-0 PMID:35798775

Buja, I., Sabella, E., Monteduro, A. G., Chiriacò, M. S., De Bellis, L., Luvisi, A., & Maruccio, G. (2021). Advances in plant disease detection and monitoring: From traditional assays to in-field diagnostics. *Sensors (Basel), 21*(6), 2129. doi:10.339021062129 PMID:33803614

Dahiya, S., Gulati, T., & Gupta, D. (2022). Performance analysis of deep learning architectures for plant leaves disease detection. Measurement. *Sensors (Basel), 24*, 100581.

Dananjayan, S., Tang, Y., Zhuang, J., Hou, C., & Luo, S. (2022). Assessment of state-of-the-art deep learning based citrus disease detection techniques using annotated optical leaf images. *Computers and Electronics in Agriculture, 193*, 106658.

Fenu, G., & Malloci, F. M. (2021). DiaMOS plant: A dataset for diagnosis and monitoring plant disease. *Agronomy (Basel), 11*(11), 2107. doi:10.3390/agronomy11112107

Filonenko, A., Kurnianggoro, L., & Jo, K. H. (2017, July). Comparative study of modern convolutional neural networks for smoke detection on image data. In *2017 10th international conference on human system interactions (HSI)* (pp. 64-68). IEEE. 10.1109/HSI.2017.8004998

Harakannanavar, S. S., Rudagi, J. M., Puranikmath, V. I., Siddiqua, A., & Pramodhini, R. (2022). Plant leaf disease detection using computer vision and machine learning algorithms. *Global Transitions Proceedings, 3*(1), 305–310. doi:10.1016/j.gltp.2022.03.016

Jadhav, V. S. (2021). Development and advancement in image processing technique for detection of various diseases in Fruit & Vegetable Plants: A review. *Int. J. Comput. Sci. Mob. Comput., 10*(2), 31–38. doi:10.47760/ijcsmc.2021.v10i02.004

Kartikeyan, P., & Shrivastava, G. (2021). Review on emerging trends in detection of plant diseases using image processing with machine learning. *International Journal of Computer Application, 975*, 8887.

Kinger, S., & Kulkarni, V. (2021, August). Explainable ai for deep learning based disease detection. In *2021 Thirteenth International Conference on Contemporary Computing (IC3-2021)* (pp. 209-216). 10.1145/3474124.3474154

Kurmi, Y., & Gangwar, S. (2022). A leaf image localization based algorithm for different crops disease classification. *Information Processing in Agriculture, 9*(3), 456–474. doi:10.1016/j.inpa.2021.03.001

Lamba, M., Gigras, Y., & Dhull, A. (2021). Classification of plant diseases using machine and deep learning. *Open Computer Science, 11*(1), 491–508. doi:10.1515/comp-2020-0122

Liu, J., & Wang, X. (2021). Plant diseases and pests detection based on deep learning: A review. *Plant Methods, 17*(1), 1–18. doi:10.118613007-021-00722-9 PMID:33627131

Marcelino, P. (2018). Transfer learning from pre-trained models. *Towards data science, 10*, 23.

Martinelli, F., Scalenghe, R., Davino, S., Panno, S., Scuderi, G., Ruisi, P., Villa, P., Stroppiana, D., Boschetti, M., Goulart, L. R., Davis, C. E., & Dandekar, A. M. (2015). Advanced methods of plant disease detection. A review. *Agronomy for Sustainable Development*, *35*(1), 1–25. doi:10.100713593-014-0246-1

Mohameth, F., Bingcai, C., & Sada, K. A. (2020). Plant disease detection with deep learning and feature extraction using plant village. *Journal of Computer and Communications*, *8*(6), 10–22. doi:10.4236/jcc.2020.86002

Nagamani, H. S. (2019). Research Methods for Plant Health Detection using Computer Vision Techniques: A Survey. *International Journal of Applied Engineering Research: IJAER*, *14*(7), 1627–1632. doi:10.37622/IJAER/14.7.2019.1627-1632

Neelakantan, P. (2021). Analyzing the best machine learning algorithm for plant disease classification. *Materials Today: Proceedings*, 2214–7853.

Peker, M. (2021). Multi-channel capsule network ensemble for plant disease detection. *SN Applied Sciences*, *3*(7), 707. doi:10.100742452-021-04694-2

Rangarajan Aravind, K., & Raja, P. (2020). Automated disease classification in (Selected) agricultural crops using transfer learning. *Automatika: časopis za automatiku, mjerenje, elektroniku, računarstvo i komunikacije, 61*(2), 260-272.

Raschka, S. (2014). An overview of general performance metrics of binary classifier systems. arXiv preprint arXiv:1410.5330.

Ravichandra, N. G. (2013). *Fundamentals of plant pathology*. PHI Learning Pvt. Ltd.

Rizzo David, M., Lichtveld, M., Mazet, J. A., Togami, E., & Miller, S. A. (2011). Plant health and its effects on food safety and security in a one health framework: Four case studies. *One Health Outlook*, *3*(6). PMID:33829143

Rout, R., & Parida, P. (2020). A review on leaf disease detection using computer vision approach. In *Intelligent Techniques and Applications in Science and Technology: Proceedings of the First International Conference on Innovations in Modern Science and Technology 1* (pp. 863-871). Springer International Publishing. 10.1007/978-3-030-42363-6_99

Ryo, M. (2022). Explainable artificial intelligence and interpretable machine learning for agricultural data analysis. *Artificial Intelligence in Agriculture*, *6*, 257–265. doi:10.1016/j.aiia.2022.11.003

Saleem, M. H., Potgieter, J., & Arif, K. M. (2019). Plant disease detection and classification by deep learning. *Plants*, *8*(11), 468. doi:10.3390/plants8110468 PMID:31683734

Shoaib, M., Hussain, T., Shah, B., Ullah, I., Shah, S. M., Ali, F., & Park, S. H. (2022). Deep learning-based segmentation and classification of leaf images for detection of tomato plant disease. *Frontiers in Plant Science*, *13*, 1031748. doi:10.3389/fpls.2022.1031748 PMID:36275583

Shurtleff, M. C., Pelczar, R. M., Kelman, A. & Pelczar, M. J. (2023, July 7). *plant disease*. Encyclopedia Britannica. https://www.britannica.com/science/plant-disease

Singh, D., Jackson, G., Hunter, D., Fullerton, R., Lebot, V., Taylor, M., Iosefa, T., Okpul, T., & Tyson, J. (2012). Taro leaf blight—A threat to food security. *Agriculture*, *2*(3), 182–203. doi:10.3390/agriculture2030182

Singh, D., Jain, N., Jain, P., Kayal, P., Kumawat, S., & Batra, N. (2020). PlantDoc: A dataset for visual plant disease detection. In *Proceedings of the 7th ACM IKDD CoDS and 25th COMAD* (pp. 249-253). ACM. 10.1145/3371158.3371196

Sinshaw, N. T., Assefa, B. G., Mohapatra, S. K., & Beyene, A. M. (2022). Applications of Computer Vision on Automatic Potato Plant Disease Detection: A Systematic Literature Review. *Computational Intelligence and Neuroscience*, *2022*, 2022. doi:10.1155/2022/7186687 PMID:36419507

Sokolova, M., Japkowicz, N., & Szpakowicz, S. (2006, December). Beyond accuracy, F-score and ROC: a family of discriminant measures for performance evaluation. In *Australasian joint conference on artificial intelligence* (pp. 1015-1021). Berlin, Heidelberg: Springer Berlin Heidelberg.

ur Rahman, H., Ch, N. J., Manzoor, S., Najeeb, F., Siddique, M. Y., & Khan, R. A. (2017). A comparative analysis of machine learning approaches for plant disease identification. *advancements in life sciences, 4*(4), 120-126.

Vishnoi, V. K., Kumar, K., & Kumar, B. (2022). A comprehensive study of feature extraction techniques for plant leaf disease detection. *Multimedia Tools and Applications*, *81*(1), 1–53. doi:10.100711042-021-11375-0

ENDNOTE

1 https://testbook.com/biology/plant-diseases

Chapter 5
Enhancing Tomato Fruit Detection and Counting Through AI-Enabled Agricultural Innovations

S Gandhimathi Alias Usha
https://orcid.org/0000-0003-1908-6249
Velammal College of Engineering and Technology, Madurai, India

ABSTRACT

This chapter aims to enhance tomato fruit detection and counting in agricultural practices through AI-enabled innovations. Traditional manual methods for fruit detection and counting are labor-intensive and time-consuming. By exploiting AI technologies, such as computer vision and machine learning algorithms, this research proposes an automated system to accurately detect and count tomato fruits in real-time. The system utilizes image processing techniques and trained models to analyze images or videos captured in the field. The proposed approach has the potential to significantly improve efficiency, reduce costs, and increase accuracy in tomato fruit detection and counting, thereby benefiting the agricultural industry.

1. INTRODUCTION

Tomato fruit plays a significant role in Indian agriculture due to its economic, nutritional, and cultural significance. Tomato is one of the most commercially important vegetable crops in India. It is widely cultivated across the country and contributes significantly to agricultural income and employment generation. Tomato

DOI: 10.4018/978-1-6684-9975-7.ch005

fruit is highly nutritious and provides essential vitamins (such as vitamin C, A, and K), minerals, and antioxidants. It is a rich source of lycopene, a powerful antioxidant associated with numerous health benefits, including reduced risk of certain types of cancer and cardiovascular diseases. Its nutritional value and versatility make it a popular ingredient in various Indian cuisines, promoting healthy eating habits. Tomato is a widely consumed vegetable in India, and its availability throughout the year is crucial for ensuring food security. Its versatility allows it to be used in a wide range of dishes, including curries, salads, soups, and sauces. The cultivation and consistent supply of tomato fruit contribute to meeting the nutritional needs of the population, enhancing food security at both the household and national levels. Tomato fruit plays a multifaceted role in Indian agriculture. Its economic importance, nutritional value, role in ensuring food security, agricultural diversity, export potential, employment generation, and contribution to soil health make it a vital crop in the Indian agricultural landscape. The cultivation and utilization of tomato fruit contribute significantly to the agricultural sector's growth, farmer livelihoods, and the overall well-being of the population (Walter et al., 2017).

The rising global population and the need for enhanced food production act as catalysts for seeking technological solutions that can optimize crop yield, reduce labor-intensive processes, and minimize resource wastage. The potential impact of AI in addressing these challenges specifically in tomato fruit detection and counting serves as a compelling motivation for this chapter.

2. OVERVIEW OF TOMATO FRUIT DETECTION AND COUNTING

Tomato fruit detection and counting play a crucial role in modern agriculture, enabling farmers to accurately assess crop yield, monitor fruit quality, and optimize harvesting processes. This section provides an overview of the significance of tomato fruit detection and counting, the traditional methods employed in the industry, their limitations, and the role of AI-enabled innovations in revolutionizing these practices.

2.1 Importance of Tomato Fruit Detection and Counting

Tomatoes are one of the most widely cultivated and economically important crops worldwide. Accurate detection and counting of tomato fruits are essential for various reasons. Firstly, it allows farmers to estimate the yield of their tomato crops, providing critical information for planning and decision-making regarding harvest timing, resource allocation, and market supply. Timely and accurate yield estimation contributes to better management of resources, reducing waste and maximizing profitability.

Secondly, tomato fruit detection and counting are crucial for quality control purposes. By accurately assessing the number of fruits and their size, shape, and color, farmers can evaluate the market readiness of their crop. This information helps in determining the optimal harvest time, ensuring that the tomatoes are picked at their peak ripeness and are of desired quality. Improved quality control enhances customer satisfaction, reduces post-harvest losses, and maintains the reputation of the farm or brand.

Furthermore, tomato fruit detection and counting enable the monitoring of crop health and the early detection of diseases or pest infestations. By identifying abnormal fruit growth patterns or discoloration, farmers can take prompt action to mitigate the spread of diseases and pests, reducing crop losses and minimizing the need for extensive chemical interventions. This proactive approach to crop monitoring and management contributes to sustainable and environmentally friendly farming practices.

2.2 Traditional Methods

Historically, tomato fruit detection and counting have relied on manual labor and visual inspection. Farm workers would physically count and collect tomatoes from the plants, which was a time-consuming and labor-intensive process. Manual methods often led to inconsistencies and errors, as human judgment can be subjective and prone to fatigue.

Additionally, visual inspection alone was insufficient for accurately detecting and counting tomato fruits, especially in large-scale farming operations. This limitation posed challenges in achieving efficient harvest planning and yield estimation. Traditional methods lacked the precision, scalability, and speed required to keep pace with the demands of modern agriculture.

2.3 Limitations of Traditional Methods

The traditional methods of tomato fruit detection and counting suffer from several limitations. Firstly, manual counting is highly dependent on the availability and reliability of skilled labor, which can be scarce and costly in some regions. The human-based approach is also vulnerable to errors, inconsistencies, and subjective interpretations, leading to inaccurate yield estimates and inefficient resource allocation.

Moreover, manual counting is time-consuming, impeding the overall productivity of the farm. As the scale of tomato farming operations increases, the labor-intensive nature of traditional methods becomes a significant bottleneck. Additionally, visual inspection alone may not capture all the fruits, especially when they are partially hidden or scattered in dense foliage, leading to underestimation of crop yield.

2.4 Role of AI-Enabled Innovations

AI-enabled innovations have the potential to revolutionize tomato fruit detection and counting, overcoming the limitations of traditional methods. Machine learning algorithms and deep learning techniques, such as convolutional neural networks (CNNs) and recurrent neural networks (RNNs), have demonstrated remarkable capabilities in image recognition and object detection tasks (Kamilaris & Prenafeta-Boldú, 2018).

By training AI models on large datasets of tomato fruit images, these algorithms can learn to accurately detect and count tomato fruits with high precision and efficiency. AI-enabled innovations leverage the power of automation, enabling real-time monitoring, rapid data processing, and scalable applications across different farming environments. They offer the potential for improved accuracy, reduced labor costs, and enhanced decision-making capabilities.

3. DATASET ACQUISITION AND ANNOTATION

The acquisition and annotation of datasets are crucial steps in developing effective AI models for tomato fruit detection and counting. This section explores the various aspects involved in dataset acquisition and annotation, including the availability of tomato fruit image datasets, ground truth annotation, challenges in dataset acquisition and annotation, and strategies for preparing high-quality datasets.

3.1 Tomato Fruit Image Datasets

To train AI models for tomato fruit detection and counting, access to diverse and representative datasets is essential. Tomato fruit image datasets consist of a collection of images that capture different varieties, growth stages, lighting conditions, and orientations of tomato fruits. These datasets serve as the foundation for training and evaluating AI models.

Fortunately, there have been efforts to curate and release publicly available tomato fruit image datasets. These datasets are typically compiled by agricultural research institutions, universities, or collaborative initiatives. They encompass a wide range of tomato varieties and contain images captured in controlled environments, greenhouses, or field conditions. These datasets serve as valuable resources for researchers and developers in the agricultural domain.

3.2 Ground Truth Annotation

To leverage the potential of tomato fruit image datasets, accurate and detailed annotation is necessary. Ground truth annotation involves manually labeling the tomato fruits in the images, providing pixel-level bounding boxes or instance-level segmentation masks to identify the precise location and extent of each fruit. This annotation process requires expertise and attention to detail.

Expert annotators carefully examine each image, identifying and outlining the boundaries of individual tomato fruits. The annotation process ensures that the AI model learns to recognize and differentiate tomatoes from other objects or background elements accurately. Ground truth annotation establishes a reference point for training the AI models, enabling them to learn the characteristics and features of tomato fruits.

3.3 Challenges In Dataset Acquisition and Annotation

Acquiring and annotating high-quality tomato fruit datasets come with several challenges. Firstly, obtaining a diverse and representative dataset that covers the full spectrum of tomato fruit variations, such as size, shape, color, and maturity stages, can be challenging. It requires collaboration with farmers, research institutions, or data-sharing initiatives to ensure the dataset's breadth and depth.

Secondly, the process of ground truth annotation is time-consuming and labor-intensive. It demands skilled annotators who possess domain expertise and are meticulous in their annotation tasks. The complexity of accurately delineating the boundaries of overlapping or occluded fruits adds to the annotation challenge.

Another challenge lies in ensuring the consistency and accuracy of annotation across the dataset. Annotators need to follow established guidelines and standards to maintain uniformity in the labeling process. Quality control measures, such as regular reviews and feedback loops, are crucial to minimize annotation errors and discrepancies.

3.4 Strategies for High-Quality Dataset Preparation

To address the challenges in dataset acquisition and annotation, several strategies can be employed to ensure the creation of high-quality datasets. Firstly, collaboration and data-sharing initiatives among multiple stakeholders, including farmers, researchers, and organizations, can enhance dataset diversity and representativeness. Sharing datasets across regions and climates allows for the inclusion of various tomato varieties and growth conditions.

Secondly, investing in training and certifying annotators with domain knowledge can improve the accuracy and consistency of ground truth annotation. Establishing clear annotation guidelines, conducting regular calibration sessions, and providing feedback mechanisms contribute to maintaining high-quality annotations.

Moreover, leveraging crowd-sourcing platforms can aid in dataset acquisition and annotation. These platforms allow a large number of contributors to participate in the annotation process, speeding up the data labeling and ensuring diverse perspectives. However, quality control measures and validation checks must be implemented to maintain annotation accuracy.

Lastly, iterative dataset refinement and augmentation techniques can enhance dataset quality (Magomadov, 2019). By incorporating feedback from AI model training results, dataset limitations and gaps can be identified and addressed.

4. TOMATO FRUIT DETECTION METHODS

Detecting tomato fruits accurately and efficiently is essential for optimizing harvest processes and yield estimation. This section explores various tomato fruit detection methods, including region-based detection techniques such as Selective Search, Faster R-CNN, and Mask R-CNN, as well as Single Shot Detection (SSD) and You Only Look Once (YOLO) approaches (Rahnemoonfar & Sheppard, 2017; Ren et al., 2020).

4.1 Region-Based Detection

Region-based detection methods have been widely employed for tomato fruit detection. These techniques involve two main stages: region proposal generation and object classification.

4.1.1 Selective Search

Selective Search is a popular region proposal algorithm that generates a diverse set of potential object regions in an image. It identifies regions based on color, texture, and intensity similarity, aiming to capture objects of varying sizes and shapes. Selective Search proposes a set of candidate regions that are then fed into a subsequent classification algorithm for object detection.

4.1.2 Faster R-CNN

Faster R-CNN (Region-based Convolutional Neural Networks) is an advanced region-based detection framework. It combines a region proposal network (RPN) with a CNN-based object classifier. The RPN generates region proposals based on anchor boxes and scores them for objectness. The selected proposals are then classified using a CNN to determine the presence of tomato fruits and their precise locations.

4.1.3 Mask R-CNN

Mask R-CNN extends Faster R-CNN by adding a mask prediction branch. In addition to detecting tomato fruits and their bounding boxes, Mask R-CNN also predicts pixel-level segmentation masks, accurately delineating the boundaries of the fruits. This fine-grained segmentation information enables more precise fruit counting and analysis.

4.2 Single Shot Detection (SSD)

Single Shot Detection (SSD) is a one-stage object detection method that directly predicts the class labels and bounding box coordinates in a single pass of the network. SSD achieves real-time performance by employing multiple feature maps at different scales to detect objects of varying sizes. It provides high-speed inference capabilities while maintaining competitive accuracy.

4.3 YOLO (You Only Look Once)

YOLO (You Only Look Once) is another popular one-stage object detection approach. YOLO divides the input image into a grid and predicts bounding boxes and class probabilities directly using a single neural network. It simultaneously predicts multiple objects in the image, allowing for fast and efficient detection. YOLO is known for its real-time performance and robustness.

5. TOMATO FRUIT COUNTING TECHNIQUES

Accurate tomato fruit counting is essential for yield estimation, harvest planning, and overall crop management. This section explores various tomato fruit counting techniques, including traditional approaches, regression-based counting, density estimation, and detection-based counting (Bargoti & Underwood, 2017; Sa et al., 2016).

5.1 Traditional Counting Approaches

Traditional tomato fruit counting methods often involve manual labor, where workers physically count and collect the fruits from plants. This approach is time-consuming, labor-intensive, and prone to errors. However, it has been a widely used method in small-scale farming operations. With advancements in technology, automated counting techniques have gained prominence (Katarzyna & Paweł, 2019).

5.2 Regression-Based Counting

Regression-based counting techniques utilize machine learning algorithms to estimate the count based on various features extracted from images. These features can include color, shape, texture, and size characteristics of the tomato fruits. Regression models, such as Support Vector Regression (SVR) or Random Forest Regression, are trained using annotated datasets to establish a relationship between the input features and the actual fruit count (Villacrés & Auat Cheein, 2020).

5.3 Density Estimation

Density estimation approaches consider the spatial distribution of tomato fruits in an image to estimate the count. These methods involve dividing the image into smaller sub regions and then calculating the density of fruit instances within each sub region. The total count is obtained by summing up the estimated densities across the entire image. Density estimation techniques, such as Kernel Density Estimation (KDE) or Gaussian Mixture Models (GMM), are commonly used for this purpose.

5.4 Detection-Based Counting

Detection-based counting methods leverage object detection algorithms to detect individual tomato fruits and count them. Techniques such as Faster R-CNN or YOLO, which were discussed in the previous section, can be adapted for fruit counting by modifying the output layer to provide count information (Chung & Van Tai, 2019). The detection algorithm identifies and localizes each fruit instance, enabling an accurate count based on the number of detected objects (Wang & Chen, 2018).

6. PROPOSED SYSTEM

6.1 Proposed Methodology

The proposed methodology is shown in Figure 1. Mask R-CNN (Region-based Convolutional Neural Network) is a popular deep learning algorithm used for object detection and instance segmentation tasks. It is an extension of the Faster R-CNN algorithm that not only detects objects but also generates pixel-level masks for each instance.

In the context of tomato fruit detection and counting, Mask R-CNN can play a significant role. Mask R-CNN can accurately detect tomato fruits within an image by identifying bounding boxes around them. The network is trained on a large dataset of annotated tomato images, learning to recognize the unique features of tomatoes. Mask R-CNN goes beyond object detection by providing pixel-level segmentation masks for each detected tomato fruit. This allows for precise identification and separation of individual tomatoes, even when they are touching or overlapping. The generated segmentation masks from Mask R-CNN can be used to count the number of tomato fruits in an image. By analyzing the individual instances and their respective masks, the algorithm can determine the total count by identifying distinct regions of the image associated with each tomato fruit (Kestur et al., 2019).

By utilizing Mask R-CNN for tomato fruit detection and counting, it becomes possible to automate these tasks, saving time and effort compared to manual counting or traditional computer vision methods. It can be particularly useful in agricultural applications, where accurate yield estimation and monitoring are crucial. Figure 2 shows the data set used in this work.

Figure 1. Proposed algorithm for tomato fruit detection

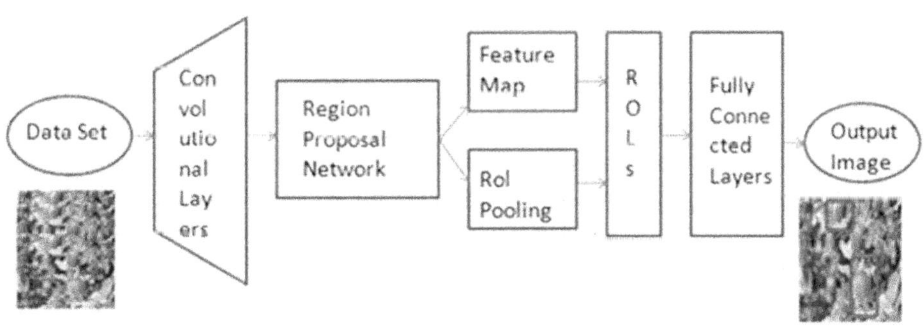

Figure 2. Data set

6.2. Experimentation Results

Evaluating the accuracy of tomato fruit counting methods requires appropriate metrics. We discuss recall, precision and the F1-score, which are the most widely and often employed evaluation metrics in the field of object identification and classification. Precision measures the ratio of correctly counted fruits to the total predicted fruits, while Recall represents the ratio of correctly counted fruits to the total actual fruits.

F1-score combines precision and recall into a single value, providing a balanced measure of counting performance (Santos et al., 2020). These metrics provide quantitative measures of the counting accuracy and allow for comparison between different counting techniques.

Mathematically, precision and recall are expressed as,

Precision = TP /(TP + FN) * 100%

Recall = TP /(TP + FN) * 100%

F1-score = 2 Precision * Recall / (Precision + Recall) * 100%

The experimentation results are shown in Figure 3. The loss function is calculated for various epochs and given in Figure 3.

The findings using MaskRCNN exhibit significantly fewer false positives and false negatives as compared to the results acquired using traditional segmentation (hand-crafted features). MaskRCNN achieves noticeably superior precision and recall values. Thus, it outperforms traditional computer vision based on colour and

Figure 3. Convergence of various loss functions in Mask-RCNN for our tomato dataset

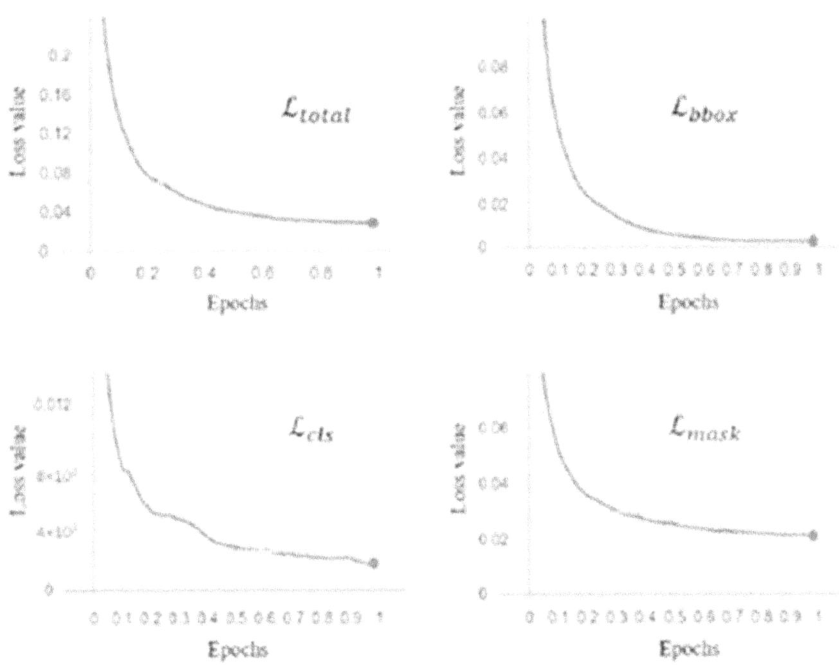

geometry in terms of handling dataset variability. Using MaskRCNN for tomato detection on our data outperforms the metrics mentioned in earlier studies. The accuracy and recall numbers for MaskRCNN are higher than the average precision of 0.82 reported by Zhou et al. (2017) and the values of precision of 0.88 and recall of 0.8 from Yamamoto et al. (2014).

It is important to consider the specific requirements and constraints of the application when selecting a counting technique and evaluating its performance. Factors such as computational efficiency, accuracy, and real-time processing capabilities play a significant role in determining the suitability of a particular method for a given scenario.

CONCLUSION

In this chapter, we explored the role of AI-enabled innovations in enhancing tomato fruit detection and counting in agricultural practices. We discussed various topics, including the importance of tomato fruit detection and counting, traditional methods and their limitations, the role of AI-enabled innovations, dataset acquisition and annotation, detection methods, counting techniques, evaluation metrics, and case studies highlighting the practical applications and impact of AI in tomato farming.

The contributions of AI-enabled innovations in tomato farming are significant. These innovations have transformed labor-intensive and error-prone processes into automated and efficient systems. They have improved productivity, reduced labor costs, minimized crop losses, and enhanced the overall quality and yield of tomatoes. AI has also promoted sustainable farming practices by optimizing resource use, reducing chemical inputs, and enabling targeted interventions. Furthermore, AI innovations have empowered farmers with real-time data, enabling them to make informed decisions based on accurate and timely information. This has led to better crop management, increased profitability, and improved decision-making processes in tomato farming.

REFERENCES

Bargoti, S., & Underwood, J. (2017). Deep fruit detection in orchards. *Proceedings of the 2017 IEEE International Conference on Robotics and Automation (ICRA).*

Chung, D. T. P., & Van Tai, D. (2019). A fruits recognition system based on a modern deep learning technique. *Journal of Physics: Conference Series*, *1327*(1), 012050. doi:10.1088/1742-6596/1327/1/012050

Kamilaris, A., & Prenafeta-Boldú, F. X. (2018). A review of the use of convolutional neural networks in agriculture. *Journal of Agricultural Science*, *156*(3), 312–322. doi:10.1017/S0021859618000436

Katarzyna, R., & Paweł, M. (2019). A vision-based method utilizing deep convolutional neural networks for fruit variety classification in uncertainty conditions of retail sales. *Applied Sciences (Basel, Switzerland)*, *9*(19), 3971. doi:10.3390/app9193971

Kestur, R., Meduri, A., & Narasipura, O. (2019). MangoNet: A deep semantic segmentation architecture for a method to detect and count mangoes in an open orchard. *Engineering Applications of Artificial Intelligence*, *77*, 59–69. doi:10.1016/j.engappai.2018.09.011

Magomadov, V. S. (2019). Deep Plearning and its role in smart agriculture. *Journal of Physics: Conference Series*, *1399*(4), 044109. doi:10.1088/1742-6596/1399/4/044109

Rahnemoonfar, M., & Sheppard, C. (2017). Deep count: Fruit counting based on deep simulated learning. *Sensors (Basel)*, *17*(4), 905. doi:10.339017040905 PMID:28425947

Ren, C., Kim, D. K., & Jeong, D. (2020). A survey of deep learning in agriculture: Techniques and their applications. *Journal of Information Processing Systems*, *16*, 1015–1033.

Sa, I., Ge, Z., Dayoub, F., Upcroft, B., Perez, T., & McCool, C. (2016). Deepfruits: A fruit detection system using deep neural networks. *Sensors (Basel)*, *16*(8), 1222. doi:10.339016081222 PMID:27527168

Santos, T. T., de Souza, L. L., dos Santos, A. A., & Avila, S. (2020). Grape detection, segmentation, and tracking using deep neural networks and three-dimensional association. *Computers and Electronics in Agriculture*, *170*, 105247. doi:10.1016/j.compag.2020.105247

Villacrés, J. F., & Auat Cheein, F. (2020). Detection and characterization of cherries: A deep learning usability case study in Chile. *Agronomy (Basel)*, *10*(6), 835. doi:10.3390/agronomy10060835

Walter, A., Finger, R., Huber, R., & Buchmann, N. (2017). Opinion: Smart farming is key to developing sustainable agriculture. *Proceedings of the National Academy of Sciences of the United States of America*, *114*(24), 6148–6150. doi:10.1073/pnas.1707462114 PMID:28611194

Wang, S. H., & Chen, Y. (2018). Fruit category classification via an eight-layer convolutional neural network with parametric rectified linear unit and dropout technique. *Multimedia Tools and Applications*, *79*(21-22), 15117–15133. doi:10.100711042-018-6661-6

Chapter 6
Harnessing Environmental Intelligence to Enhance Crop Management by Leveraging Deep Learning Technique

S. Gandhimathi Alias Usha
https://orcid.org/0000-0003-1908-6249
Velammal College of Engineering and Technology, Madurai, India

ABSTRACT

This chapter aims to enhance crop management practices by harnessing environmental intelligence through the power of deep learning techniques. Efficient and sustainable crop management is crucial for meeting the increasing demand for agricultural products while minimizing environmental impact. In recent years, the integration of deep learning techniques with environmental data has shown great potential in improving crop management practices. The proposed approach involves training deep learning model Dense Net 121 to predict important crop management factors, including yield estimation, disease, and pest outbreaks. The models are trained using historical and real-time data, enabling them to adapt and respond to dynamic environmental conditions. By capturing complex patterns and interactions, deep learning models can provide valuable insights and recommendations to farmers, enabling them to optimize resource allocation, reduce input wastage, and improve overall crop productivity.

1. INTRODUCTION

Crop management plays a pivotal role in modern agriculture, encompassing a range of practices and strategies aimed at optimizing crop productivity, minimizing

DOI: 10.4018/978-1-6684-9975-7.ch006

environmental impact, and ensuring long-term sustainability. With the world's population steadily increasing, crop management practices have become even more critical to meet global food demands while safeguarding natural resources. This chapter explores the importance of crop management in agriculture and highlights key aspects and strategies involved in its implementation (Alston & Pardey, 2014).

Maintaining soil health and fertility is a fundamental component of effective crop management. Soil assessments, including analysis of nutrient content, pH levels, and organic matter, provide valuable insights into the specific needs of crops. By adopting practices like crop rotation, cover cropping, and organic matter addition, farmers can enhance soil structure, nutrient availability, and microbial activity. Improved soil health not only boosts crop yields, but also reduces erosion, enhances water retention, and promotes overall ecosystem resilience. Efficient water management is imperative for sustainable crop production. Crop management practices such as precision irrigation techniques, soil moisture monitoring, and water-use efficiency strategies help minimize water wastage, conserve water resources, and mitigate the risk of water scarcity. By matching irrigation schedules to crop water requirements and employing technologies like drip irrigation or sensor-based systems, farmers can optimize water usage while ensuring adequate hydration for crops.

Pests and diseases pose significant threats to crop health and productivity. Integrated Pest Management (IPM) practices are essential for effective crop management. IPM combines various techniques, including biological control (such as beneficial insects), cultural practices (like crop rotation and trap crops), and targeted pesticide applications, to manage pest populations while minimizing the use of chemical pesticides. By integrating pest's and disease monitoring, prevention, and control measures, farmers can reduce crop losses, preserve beneficial organisms, and safeguard the environment (Li et al.,). Proper nutrient management is crucial for optimal crop growth and productivity. Soil testing and analysis help determine the precise nutrient requirements of crops. By implementing balanced fertilization practices, including the judicious use of organic fertilizers, farmers can supply essential nutrients in the right amounts and at the right time. Precision application techniques, such as variable rate fertilization, ensure optimal nutrient distribution across fields, minimizing nutrient runoff and pollution while maximizing crop uptake. Weeds compete with crops for resources and can significantly reduce yields. Effective weed control is vital for crop management. Integrated weed management practices, combining mechanical methods (such as tillage or mulching), cultural practices (like crop rotation), and targeted herbicide use, help suppress weed growth and minimize their impact on crop productivity. Selective herbicides are preferred to minimize damage to desirable plants and reduce chemical input.

Crop management today increasingly relies on data-driven decision making. Modern technologies, such as remote sensing, drones, and automated sensor systems, provide real-time and high-resolution data on crop health, growth patterns, and environmental conditions. By integrating data from various sources and employing advanced analytics, farmers gain valuable insights for precise management decisions related to planting times, irrigation scheduling, fertilizer application rates, and pest control strategies. This data-driven approach enhances efficiency, resource optimization, and productivity while minimizing unnecessary inputs and reducing environmental impact.

Sustainability lies at the core of crop management. Farmers who adopt sustainable practices strive to minimize adverse environmental effects and foster the long-term health of ecosystems. This includes reducing chemical inputs, conserving water resources, preserving biodiversity, and adopting climate-smart agricultural techniques. Sustainable crop management practices ensure the preservation of natural resources and contribute to resilient agricultural systems capable of withstanding environmental challenges .Hence, Crop management is essential for optimizing agricultural practices, increasing productivity, reducing environmental impact, and ensuring the long-term sustainability and profitability of farming operations.

2.TRADITIONAL METHODS ADOPTED FOR CROP MANAGEMENT

Traditional methods of crop management refer to agricultural practices that have been used for generations and are based on local knowledge and experience. These methods often rely on manual labor and limited technological interventions. Crop rotation involves the systematic alternation of different crops in a specific sequence over time. This practice helps prevent the buildup of pests, diseases, and nutrient depletion in the soil. By rotating crops, farmers can maintain soil fertility, control weeds, and reduce the reliance on chemical inputs. Intercropping is the practice of growing two or more crops simultaneously in the same field. Different crops with complementary growth patterns and nutrient requirements are planted together, maximizing the use of available space, sunlight, and soil nutrients. Intercropping provides benefits such as enhanced pest control, improved soil structure, and increased biodiversity.

Companion planting involves the deliberate pairing of compatible plant species in close proximity. Certain plants have natural properties that repel pests, attract beneficial insects, or enhance soil fertility. By interplanting these companion plants, farmers can deter pests, improve pollination, and enhance overall crop health. Farmers have long practiced the art of saving and storing seeds from their own crops for future planting. This method preserves local seed varieties, which are adapted to specific environmental conditions and often possess desirable traits such as disease resistance and high yield potential. Traditional seed saving ensures seed sovereignty and reduces dependence on commercially produced seeds. Traditionally, weeds were managed through manual labor, involving practices like hand-weeding, hoeing, or tilling. Farmers would physically remove weeds from the fields to reduce competition for resources and prevent weed seed dispersal. Although labor-intensive, manual weed control is effective in maintaining crop productivity without relying heavily on herbicides.

Traditional crop management often involves the use of natural pest control methods. Farmers may introduce beneficial insects, such as ladybugs or predatory wasps, to control pests. Additionally, cultural practices like crop diversification, trap cropping, and the use of natural repellents, such as neem or garlic, can deter pests without resorting to synthetic pesticides. In regions where water resources are limited, traditional irrigation methods such as flood irrigation, furrow irrigation, or drip irrigation are employed. These techniques rely on the application of water directly to the crop root zone, minimizing water wastage through evaporation or runoff. Traditional crop management incorporates local knowledge and practices passed down through generations. Indigenous communities have developed unique farming techniques suited to their specific environments, considering factors like climate, soil types, and cultural traditions. These practices often emphasize the sustainable use of resources and promote the preservation of biodiversity.

While modern agricultural practices and technologies have evolved, traditional methods of crop management still hold value, particularly in terms of sustainability, biodiversity preservation, and cultural significance. The integration of traditional knowledge with modern scientific advancements can lead to innovative and holistic approaches to crop management.

3. ROLE OF TECHNOLOGY IN CROP MANAGEMENT

Technology plays a crucial role in modern crop management, revolutionizing traditional farming practices and enabling more efficient, precise, and sustainable agricultural systems. Here are some key roles of technology in crop management:

Precision Agriculture: Farmers can utilize technology like GPS (Global Positioning System) and GIS (Geographic Information System) to accurately map their fields and analyze multiple factors like soil attributes, nutrient levels, and yield potential. This valuable information enables farmers to make informed choices regarding seed planting, fertilization, irrigation, and pest management, thereby optimizing resource utilization and maximizing crop yields.

Remote Sensing and Drones: Remote sensing technologies, including satellite imagery and aerial drones, provide farmers with real-time and high-resolution data on crop health, growth patterns, and stress conditions (Vanbrabant et al., 2020). By analyzing this data, farmers can identify areas of concern, monitor crop performance, and take timely action to address issues such as nutrient deficiencies, diseases, or water stress.

Sensor Technology: Advanced sensors, such as moisture sensors, weather stations, and nutrient sensors, are used to monitor soil conditions, plant health, and environmental parameters. These sensors provide accurate and continuous data, enabling farmers to adjust irrigation schedules, fertilizer applications, and other management practices based on real-time conditions, improving resource efficiency and crop performance.

Automated Irrigation Systems: Technology allows for the implementation of automated irrigation systems that precisely deliver water to crops based on their specific needs. These systems can be programmed to consider factors such as soil moisture levels, weather conditions, and crop growth stages, ensuring optimal water use and minimizing water wastage (Dawod & Dobre, 2022).

Data Analytics and Farm Management Software: Advanced data analytics and farm management software enable farmers to collect, store, analyze, and interpret vast amounts of data related to crop management. By integrating data from various sources, farmers can gain valuable insights into crop performance, identify trends, and make informed decisions to improve efficiency, productivity, and sustainability.

Crop Monitoring and Decision Support Systems: Technology aids in the development of decision support systems that provide real-time information and recommendations for crop management. By incorporating data inputs like weather forecasts, soil moisture levels, pest and disease models, as well as historical data, these systems support farmers in making well-informed decisions regarding planting, irrigation, fertilization, and pest control.

Robotics and Automation: Robotics and automation technologies are increasingly being used in crop management tasks. Autonomous robots can perform activities such as planting, harvesting, and weed control with precision and efficiency. This reduces labor requirements, increases productivity, and reduces the environmental impact associated with manual labor and chemical inputs.

Mobile Applications: Mobile applications provide farmers with instant access to valuable information, resources, and tools related to crop management. These apps offer features such as weather updates, pest and disease identification, market prices, and farming tips, empowering farmers to make informed decisions and manage their crops more effectively.

The incorporation of technology into crop management optimizes efficiency, boosts productivity, and promotes sustainability in agricultural practices. It enables farmers to optimize resource use, reduce environmental impact, make data-driven decisions, and adapt to changing conditions, ultimately leading to improved crop yields and profitability.

4. DEEP LEARNING CONCEPTS AND ARCHITECTURES

A branch of machine learning called deep learning focuses on teaching artificial neural networks to carry out difficult tasks by emulating the structure and operation of the human brain. Many sectors, including computer vision, natural language processing, speech recognition, and autonomous systems, have been transformed by this quickly developing discipline. Artificial neural networks form the foundation of deep learning, comprising interconnected layers of artificial neurons or nodes. These nodes receive input signals, perform computations, and generate output signals. The networks are organized into layers, with each layer progressively extracting more abstract features from the input data. The final layer produces the desired output, such as a classification or prediction. Deep learning excels in handling high-dimensional and unstructured data, such as images, audio, and text, by automatically learning hierarchical representations from raw data. This ability to extract complex features without explicit feature engineering is a significant advantage of deep learning over the traditional machine learning approaches.

Neural Networks: Neural networks serve as the fundamental components of deep learning, functioning as interconnected nodes, also known as artificial neurons that are arranged into layers. Each neuron performs a mathematical operation on its input and transmits the result to the subsequent layer. The layers closer to the input are responsible for capturing basic or low-level features, while deeper layers specialize in learning more intricate and abstract representations.

Convolutional Neural Networks (CNNs): CNNs are particularly effective for analyzing visual data. They employ specialized layers, such as convolutional and pooling layers, to automatically learn spatial hierarchies of features. Convolutional layers apply filters to the input data, detecting visual patterns at different scales, whereas Pooling layers down sample the data, reducing the spatial dimensionality while retaining important information (Mohsin Kabir et al., 2020). In a variety of tasks, including picture categorization, object recognition, and image synthesis, CNNs have achieved outstanding results.

Recurrent Neural Networks (RNNs): These are specifically crafted to handle sequential data, including time series or text. With their feedback connections, RNNs possess the ability to retain and utilize information from previous time steps. This recurrent structure empowers RNNs to capture temporal relationships and glean insights from contextual cues. Long Short-Term Memory (LSTM) and Gated Recurrent Unit (GRU) represent notable advancements in RNN architecture, effectively addressing challenges like the vanishing gradient problem and facilitating the learning of long-term dependencies.

Generative Adversarial Networks (GANs): The architecture of a GAN comprises two key components: a generator and a discriminator. The generator's primary role is to produce synthetic data, such as images, that closely resemble real data. Meanwhile, the discriminator is responsible for distinguishing between real and fake data, aiming to accurately identify the generated data. Through an adversarial training process, these two networks work in opposition to each other, with the generator attempting to deceive the discriminator and the discriminator striving to accurately classify the data. This interplay between the generator and discriminator promotes the development of highly realistic synthetic data. By engaging in an adversarial training process, GANs progressively enhance their ability to produce output that is highly realistic. GANs have played a pivotal role in the generation of synthetic images, augmenting datasets through data synthesis, and propelling advancements in unsupervised learning methods.

Deep learning excels at learning intricate patterns and representations from large and complex datasets. Deep learning possesses the capability to autonomously extract pertinent features, minimizing the necessity for manual feature engineering. This attribute empowers deep learning models to effectively process a wide range of data types, encompassing images, audio, text, and time-series data. The ability to parallelize computations makes deep learning models scalable and capable of leveraging GPUs and distributed systems (Hasan et al., 2020). Transfer learning is a technique that enables pre-trained models to be fine-tuned on new tasks, thereby mitigating the requirement for large amounts of training data. By leveraging knowledge gained from previous tasks, pre-trained models can be adapted to new domains or tasks with relatively limited labeled data. This approach helps accelerate the training process, improves generalization, and enables effective utilization of existing model architectures and learned representations. Transfer learning has proven to be a valuable strategy in various domains, allowing for efficient and effective utilization of deep learning models in real-world applications (Barbedo, 2018). Deep learning models are computationally intensive and require substantial computational resources. Large amounts of labeled data are typically needed to train deep learning models effectively. Interpreting and explaining the decision-making process of deep learning models can be challenging, leading to concerns about transparency and fairness.

5. DEEP LEARNING IN CROP IDENTIFICATION

Now a day, deep learning techniques have emerged as powerful tools for image-based crop identification, revolutionizing the way farmers and researchers classify and monitor crops. Indeed, thanks to advancements in computer vision and the availability of large annotated datasets, deep learning models have exhibited remarkable accuracy in recognizing and distinguishing various crop types from images. By training on extensive datasets that contain labeled examples of different crop types, deep learning models can learn intricate visual patterns and features associated with each crop. This enables them to accurately classify and differentiate between different crops based on visual characteristics captured in images (Hosseiny et al., 2020). The application of deep learning in crop recognition has proven to be valuable in agriculture, facilitating tasks such as crop monitoring, disease detection, and yield estimation.

5.1. Image-Based Crop Recognition

Image-based crop recognition involves the use of images captured by various means, such as drones, satellites, or ground-based sensors, to identify and classify different crop types. It eliminates the need for manual field surveys or time-consuming visual inspections, enabling rapid and efficient crop monitoring on a large scale. Image-based crop recognition has significant implications for precision agriculture, yield estimation, disease detection, and resource management.

5.2. Deep Learning Models for Crop Identification

CNNs have demonstrated exceptional effectiveness in crop identification tasks (Schmidhuber, 2015). CNNs are specifically tailored to extract pertinent features from images, employing a hierarchical approach that emulates the human visual system's information processing. While the pooling layers in CNNs down sample the representations to concentrate on the most important components, the convolutional layers employ filters to capture local patterns and features. This hierarchical feature extraction enables CNNs to learn and differentiate between various crop types based on their visual characteristics. The use of CNNs in crop identification has significantly advanced the field of precision agriculture and has been instrumental in automating crop monitoring and management processes.

5.2.1 Training Data and Annotation

Deep learning models necessitate a substantial quantity of labeled training data to acquire the ability to discern and generalize patterns effectively. Crop identification datasets encompass images of various crop species, growth stages, and environmental conditions. Annotation of these datasets involves labeling each image with the corresponding crop type, which can be done manually or using semi-automated methods.

5.2.2 Model Architecture and Training

Convolutional, pooling, and fully linked layers are commonly used in several layers in deep learning models for crop identification. These layers learn to extract features of varying complexity from input images. The models are trained using large-scale datasets, employing optimization algorithms to minimize the difference between predicted and actual crop labels. Transfer learning, where pre-trained models are fine-tuned on crop-specific datasets, can also enhance the performance of crop identification models.

5.2.3 Feature Extraction and Representation

Deep learning models automatically learn to extract high-level features from images, enabling them to capture unique visual characteristics of different crop types. These models can detect shape, color, texture, and spatial patterns specific to each crop, allowing for accurate identification even in complex field conditions.

5.2.4 Model Evaluation and Validation

The performance of deep learning models for crop identification is assessed through metrics such as accuracy, precision, recall, and F1-score. Models are validated using separate test datasets or through cross-validation techniques to ensure their generalizability and robustness across different datasets.

Deep learning models enable accurate and efficient crop monitoring on a large scale, providing real-time insights into crop growth, development, and health. This information helps farmers make timely decisions regarding irrigation, fertilizer application, disease control, and yield estimation. Crop identification using deep learning contributes to precision agriculture by facilitating targeted interventions. Farmers can precisely apply inputs such as water, nutrients, and pesticides based on specific crop types, reducing costs, minimizing environmental impact, and optimizing resource utilization (Chen et al., 2022). Deep learning models can identify crop diseases and pests based on visual symptoms captured in images. This early detection enables proactive management strategies, preventing the spread of diseases and minimizing yield losses. It helps farmers implement timely and targeted interventions, reducing the reliance on broad-spectrum pesticides (Mohanty et al., 2016). Accurate crop identification using deep learning models allows for precise yield estimation.

6. DEEP LEARNING BASED PLANT DISEASE DETECTION

Crop management and plant disease detection are closely related aspects of agricultural practices. Effective crop management strategies play a vital role in preventing and managing plant diseases. Timely detection of plant diseases is crucial for effective disease management. Crop management practices, such as regular field inspections and monitoring, can help identify the early signs and symptoms of diseases. Farmers and agricultural professionals trained in disease recognition can closely observe plants for any abnormal growth patterns, discoloration, lesions, or other indicators of diseases. The proposed structure for plant disease detection is shown in Figure 1.

Figure 1. Workflow diagram

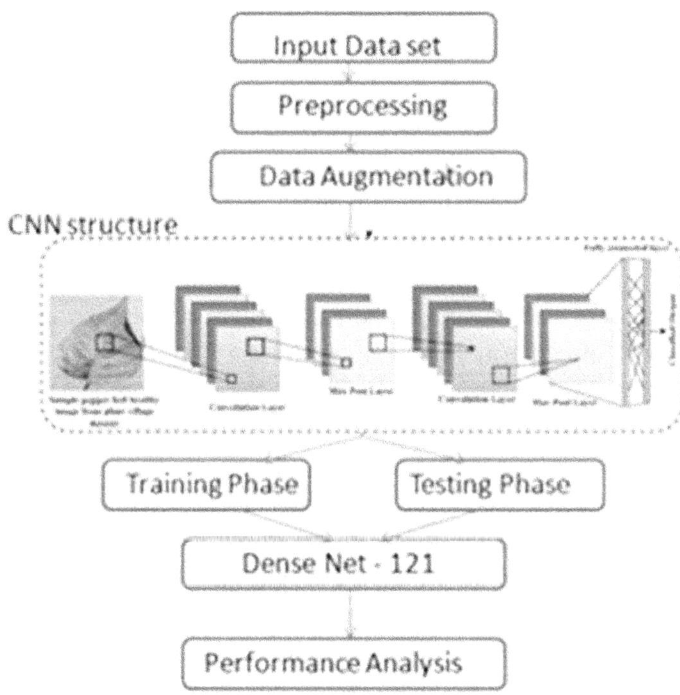

The Plant Village dataset, sourced from Plant Village (PlantVillage, n.d.), is a publicly accessible dataset that contains various categories of plant diseases, which is given in Figure 2. It consists of a collection of 54,305 images, divided into 38 distinct classes. We divided the dataset into three distinct subgroups for our experimental analysis: training samples, testing samples, and validation samples. We designated 80% of the Plant Village dataset for the training set in order to provide a sizeable amount of data for training the pre-trained models. This allowed us to leverage a sufficiently large training set to optimize the performance and generalization of the models. The remaining portions of the dataset were used for testing the models' performance on unseen data and validating their accuracy and robustness. The remaining 20% of the dataset was split between the validation set and the testing set. This division allowed us to evaluate the performance of the models on unseen data and validate their effectiveness. Among the total 54,305 samples available in the Plant Village dataset, we allocated 43,955 samples for training, 4,902 samples for validation, and 5,488 samples for testing. Importantly, all three sets (training, testing, and validation) encompassed all 38 classes, covering the various plant diseases present in the dataset. By employing this dataset and the specified data split, we aimed to train and evaluate the accuracy of the pre-trained models in accurately identifying and classifying the different plant diseases represented in the Plant Village dataset. This comprehensive dataset provided us with a diverse range of plant disease images to ensure robust training and testing of our models.

DenseNet-121 is a deep convolutional neural network model that is extensively employed for image classification tasks. The Architecture of Dense Net is shown in Fig.3. It introduces the concept of dense connectivity, wherein each layer receives input not only from its preceding layers but also from all preceding layers. This is accomplished through concatenating the feature maps generated by the previous layers and providing them as input to the subsequent layer. By incorporating these dense connections, DenseNet-121 allows for improved information flow and knowledge sharing throughout the network. The shorter connections facilitate the propagation of gradients, address the vanishing gradient problem, and encourage feature reuse, resulting in highly effective and efficient feature learning. This collective knowledge transfer ensures that each layer has access to a wide range of information from earlier layers, enabling the network to learn rich and diverse representations. To maintain efficiency and compactness, the number of channels within a dense block, which refers to a group of densely connected layers, is typically reduced. The growth rate, denoted as k, determines the number of new features or channels added at each layer within the dense block. By controlling the growth rate, the network can balance the trade-off between the complexity of feature extraction and computational efficiency. A lower growth rate result in a more compact model with fewer parameters, while higher growth rates allow the model to capture more diverse and complex features but

Figure 2. Sample images from plant village dataset for 38 types of leaf diseases

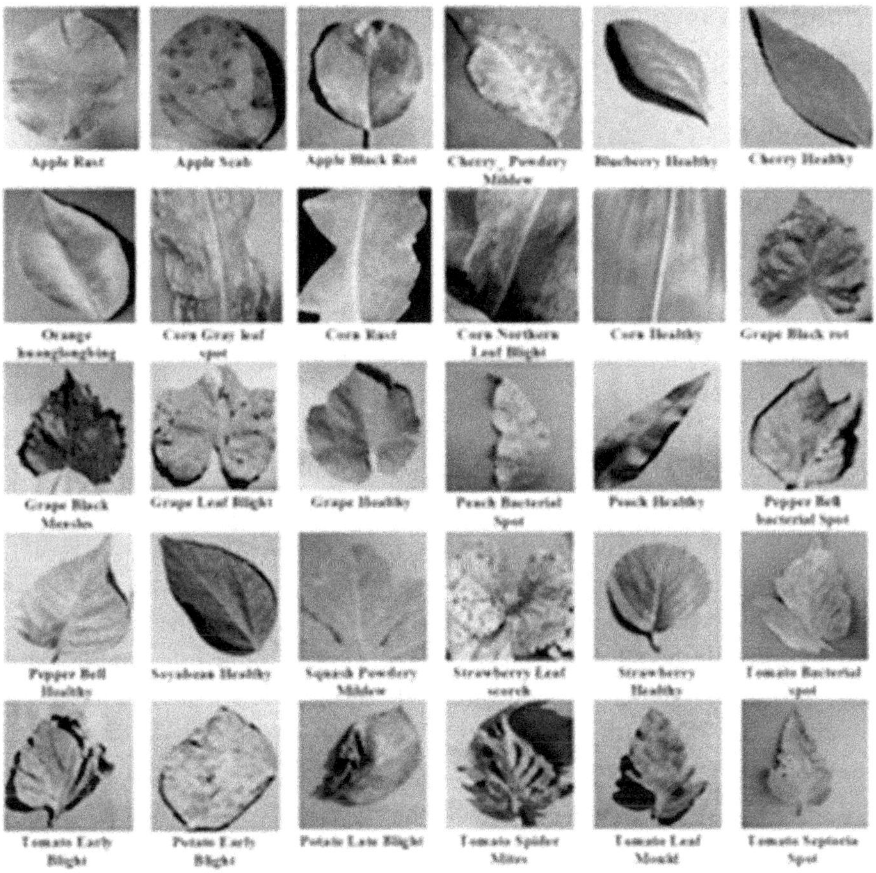

at the expense of increased computational requirements. The growth rate parameter in DenseNet-121 plays a crucial role in finding the right balance between model size and representational capacity. This design reduces the overall number of parameters and computational complexity while maintaining strong representational power.

The dense block in DenseNet-121 follows a specific sequence of operations. Within the dense block, each composition layer undergoes a series of steps. Regularization techniques are first applied to the feature maps, followed by activation functions like ReLU and convolution operations. These operations help in refining and transforming the feature maps to capture meaningful patterns and structures. In addition, batch normalization, ReLU activation, convolution, and pooling operations are applied to the output of subsequent layers to further enhance the extracted features and downscale the feature maps. It exemplifies a deep learning architecture that maximizes information flow and leverages dense connections to improve image classification performance. The network architecture is specifically engineered to be efficient and compact while enabling the acquisition of intricate representations. This can be attained by knowledge sharing between layers, as denoted by Equation 1.

$$Y= W_3\{x,h_1(x),h_2(x),h_3(x)\} \quad (1)$$

7.EXPERIMENTATION

In this study, we assessed the training and testing accuracy for each experiment conducted. During both the testing and training phases of our study, we computed the losses for the different models that we employed. Loss functions are employed to measure the difference between the model's anticipated output and the actual target values. By calculating the losses, we were able to assess the performance and evaluate the accuracy of the models. This allowed us to compare the effectiveness of different models and make informed decisions regarding their suitability for the specific task or problem at hand. Our focus was on enhancing the learning speed of the CNN through the use of transfer learning models trained on the Plant Village dataset. To achieve this, we employed pre-trained models namely DenseNet-121. This model had previously undergone training on the Image Net dataset, which consists of 1.2 million images categorized into 1,000 classes. By utilizing these pre-trained models, we aimed to leverage the knowledge and features learned from the vast Image Net dataset. This approach enables CNN to accelerate its learning process and potentially improve the accuracy of crop identification tasks on the Plant Village dataset.

Figure 3. Architecture of dense net

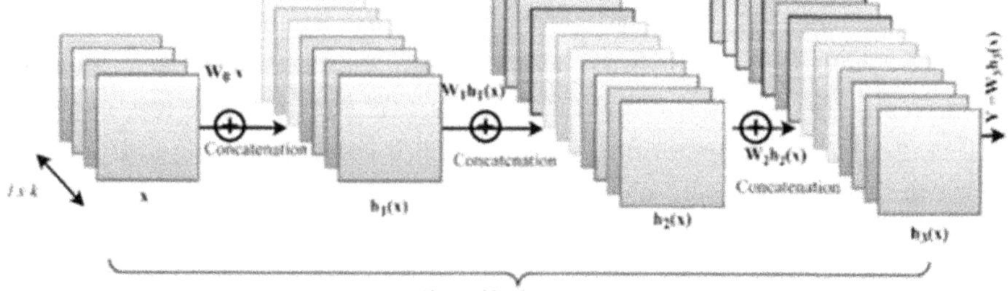

During our experiments, we continuously monitored and recorded both the training accuracy and testing accuracy of the models. These accuracy metrics were used to evaluate the performance of the models and assess their ability to correctly classify and identify crops. Additionally, we calculated the losses during both the testing and training phases, as these serve as indicators of the model's capability to minimize errors and improve its predictive accuracy. By leveraging transfer learning and utilizing pre-trained models, our objective was to enhance the efficiency and effectiveness of our crop identification system. This approach allowed us to leverage the knowledge and learned representations from pre-existing models, reducing the need for extensive training from scratch. The ultimate aim of our work was to contribute to more accurate and efficient agricultural practices through the development of an advanced crop identification system.

DenseNet-121 is designed to address the vanishing gradient problem and increase the depth of the network. It achieves this by incorporating dense blocks, each consisting of multiple layers, within the network. The details of pretrained network architecture model is given in table.1It comprises four dense blocks. Within each dense block, a combination of convolution operations using different filter sizes is repeated a specific number of times. These repetitions help capture and extract intricate features from the input data.

In the initial dense block, convolutions are conducted using filter sizes of 1x1 and 3x3, which are repeated six times. This choice enables the network to extract both detailed and broader features from the input data. Likewise, in the second dense block, convolutions employ filter sizes of 3x3 and 1x1, repeated 12 times, enhancing the depth and complexity of the learned representations. The third dense block follows a similar pattern, with convolution operations using the same filter size repeated 24 times. This repetitive process further enriches the depth and complexity of the learned representations. This increased repetition allows the network to capture increasingly complex patterns and structures. Finally, the fourth dense block repeats the steps 16 times, further deepening the network and

Table 1. Pretrained network architecture model

Parameters	Values
Total layers	121
Max Pooling layers	4
Dropout layers	-
Dense layers	4
Filter size	3x3,1x1
Stride	2x2

extracting more sophisticated features. Between the dense blocks, transition blocks are inserted. These transition blocks typically consist of convolutional and pooling layers. They serve to downscale the feature maps and reduce their dimensionality, facilitating information flow and computational efficiency.DenseNet-121 tackles the vanishing gradient problem by introducing dense blocks with repeated convolution operations. These dense blocks, with varying filter sizes and repetition patterns, enable the network to learn and extract progressively more complex features. The inclusion of transition blocks between dense blocks ensures smooth information flow and aids in down sampling the feature maps.

To address the vanishing gradient problem that can occur due to skip connections in deep neural networks, we employed regularization techniques such as batch normalization. The term "vanishing gradient problem" alludes to the difficulty in successfully updating the weights during training since the gradients grow incredibly small as they propagate backward through numerous layers. During training, a method called batch normalization normalizes the inputs inside each mini-batch. By normalizing the inputs, it helps mitigate the vanishing gradient problem and ensures more stable and efficient training. Batch normalization also acts as a regularizer by reducing internal covariate shift and improving the overall generalization ability of the model.

By applying batch normalization to the network, we aimed to stabilize the gradients flowing through the skip connections and alleviate the vanishing gradient problem, thus enabling more effective training and improving the overall performance of the model. Nevertheless, the utilization of deeper models presented certain challenges, including over fitting, covariant shifts, and increased training time complexity. In our experimental setup, we addressed these challenges by fine-tuning the hyper parameters. We specifically focused on optimizing the sensitivity metric, which measures the accuracy of correctly identifying healthy plants (true positives) and the identification of healthy plants as unhealthy (false negatives). Specificity measures the accuracy of predicting actually unhealthy plants as unhealthy (true negative) and actually unhealthy leaves as healthy (false positive). The ratio of true negatives to the total of true negatives and false positives is used to compute the specificity, also known as the true negative rate. It is an indicator of how many genuine negative cases the model accurately identified.

The fraction of true positive predictions out of all real positive cases in the dataset is measured by sensitivity, often known as the true positive rate. It is determined as the ratio of true positives to the total of true positives and false negatives, and it shows the model's capacity to detect all occurrences of positivity.

The F1 score combines precision and recall into a single metric by taking their harmonic mean. It offers a fair evaluation of the model's effectiveness by taking into account both false positives and erroneous negatives. A higher value indicates

Figure 4. a) Training and testing accuracy; b) training and validation loss in Dense Net- 121

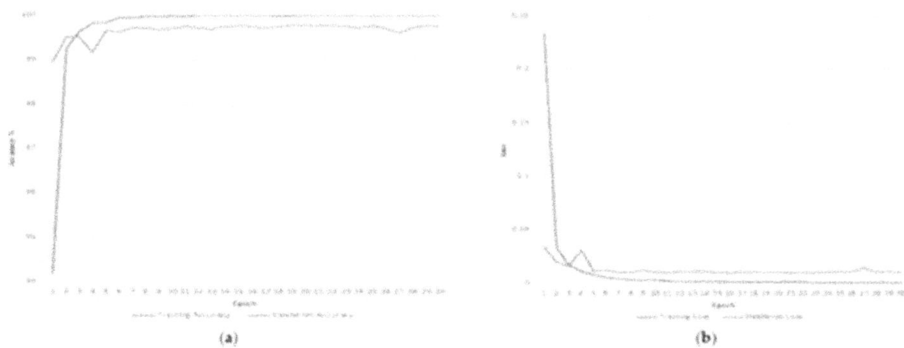

Figure 5. F1 score vs. validation scores of DenseNet-121

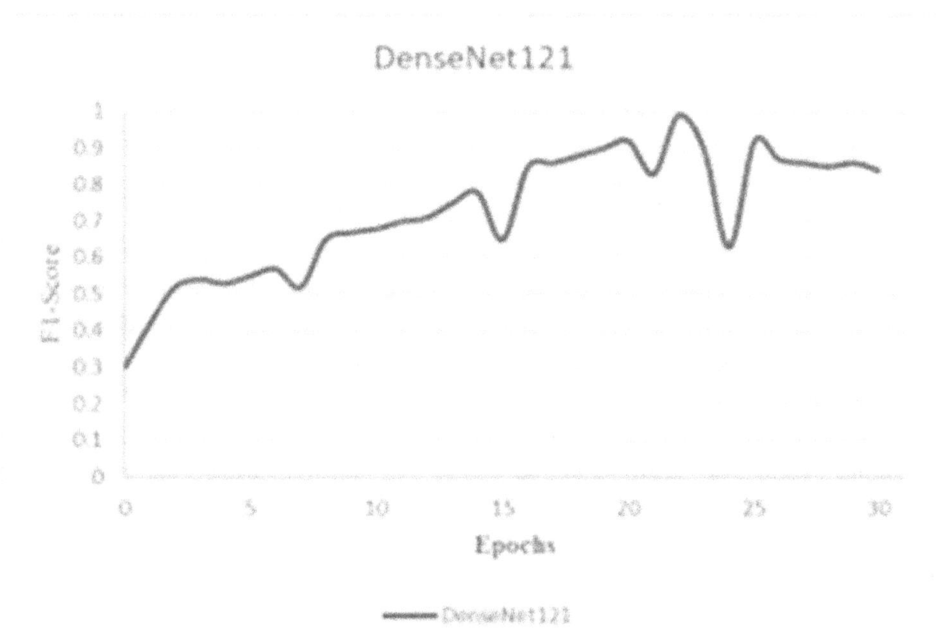

greater performance in terms of both recall and precision. The F1 score ranges from 0 to 1. When false positives and false negatives have differing effects or costs, the F1 score is especially helpful. The F1 score offers a thorough assessment of a model's performance in classification tasks by taking into account both precision and recall.

Hence, the evaluation and analysis demonstrate that DenseNet-121 models showed better performance compared other models, as indicated by their higher sensitivity, specificity, and F1 score values. Figures 4 and 5 show the performance of the system.

8. CONCLUSION

Harnessing environmental intelligence through the utilization of deep learning techniques has shown promising results in enhancing crop management practices. Deep learning algorithms have demonstrated the ability to analyze complex environmental data, such as satellite imagery, weather patterns, and soil conditions, to provide valuable insights and decision-making support for farmers and agricultural professionals. By exploiting deep learning, crop management strategies can be optimized, leading to improved resource allocation, increased yield, and sustainable farming practices.

Plant disease detection plays a crucial role in ensuring crop health, optimizing resource allocation, and minimizing yield losses in agriculture. The effectiveness and efficiency of disease detection systems have been greatly enhanced over time by technological and methodological breakthroughs. From traditional visual inspections to sophisticated sensor-based technologies and data-driven approaches, plant disease detection has witnessed remarkable progress. Indeed, the advent of artificial intelligence (AI) and machine learning algorithms has brought about significant advancements in automating disease detection processes. These technologies have demonstrated great potential in various fields, including healthcare and medical diagnostics. Through the integration of deep learning techniques, several benefits have been observed. Firstly, the accurate and timely analysis of environmental data allows for proactive decision making, enabling farmers to respond effectively to changing conditions and mitigate potential risks. This includes optimizing irrigation schedules, adjusting fertilizer applications, and implementing pest and disease management strategies. Secondly, deep learning models have shown the ability to detect subtle patterns and anomalies in environmental data, facilitating the early detection of crop's stress, disease outbreaks, and nutrient deficiencies. By identifying these issues at an early stage, appropriate interventions can be implemented promptly, minimizing crop losses and improving overall plant health.

Moreover, the utilization of deep learning techniques enables the creation of predictive models that forecast crop yields, pest and disease prevalence, and market conditions. These predictive capabilities provide farmers with valuable information for planning and decision making, enhancing operational efficiency and profitability.

REFERENCES

Alston, J. M., & Pardey, P. G. (2014). Agriculture in the Global Economy. *The Journal of Economic Perspectives*, *28*(1), 121–146. doi:10.1257/jep.28.1.121

Barbedo, J. G. A. (2018). Impact of dataset size and variety on the effectiveness of deep learning and transfer learning for plant disease classification. *Computers and Electronics in Agriculture*, *153*, 46–53. doi:10.1016/j.compag.2018.08.013

Chen, Z., Wu, R., Lin, Y., Li, C., Chen, S., Yuan, Z., & Zou, X. (2022). Plant Disease Recognition Model Based on Improved YOLOv5. *Agronomy (Basel)*, *12*(2), 365. doi:10.3390/agronomy12020365

Dawod, R. G., & Dobre, C. (2022). Upper and Lower Leaf Side Detection with Machine Learning Methods. *Sensors (Basel)*, *22*(7), 2696. doi:10.339022072696 PMID:35408307

Hasan, R. I., Yusuf, S. M., & Alzubaidi, L. (2020). Review of the state of the art of deep learning for plant diseases: A broad analysis and discussion. *Plants*, *9*(10), 1302. doi:10.3390/plants9101302 PMID:33019765

Hosseiny, B., Rastiveis, H., & Homayouni, S. (2020). An Automated Framework for Plant Detection Based on Deep Simulated Learning from Drone Imagery. *Remote Sensing (Basel)*, *12*(21), 3521. doi:10.3390/rs12213521

Li, L., Zhang, S., & Wang, B. (2021). Plant Disease Detection and Classification by Deep Learning—A Review. *IEEE Access, 9*, 56683–56698.

Mohanty, S. P., Hughes, D. P., & Salathé, M. (2016). Using Deep Learning for Image-Based Plant Disease Detection. *Frontiers in Plant Science*, *7*, 1419. doi:10.3389/fpls.2016.01419 PMID:27713752

Mohsin Kabir, M., Quwsar Ohi, A., & Mridha, M. F. A Multi-plant disease diagnosis method using convolutional neural network. arXiv 2020, arXiv:2011.05151.

PlantVillage. (n.d.). *About.* Plant Village. https://plantvillage.psu.edu/

Schmidhuber, J. (2015). Deep learning in neural networks: An overview. *Neural Networks*, *61*, 85–117. doi:10.1016/j.neunet.2014.09.003 PMID:25462637

Vanbrabant, Y., Delalieux, S., Tits, L., Pauly, K., Vandermaesen, J., & Somers, B. (2020). Pear flower cluster quantification using RGB drone imagery. *Agronomy (Basel)*, *10*(3), 407. doi:10.3390/agronomy10030407

Chapter 7
Crop Prediction for Smart Agriculture Using Ensemble of Classifiers

Khushal Kindra
Vellore Institute of Technology, Chennai, India

Bhuvaneswari Amma N. G.
https://orcid.org/0000-0003-3660-380X
Vellore Institute of Technology, Chennai, India

ABSTRACT

Nowadays due to the advancement in technology, smart agriculture is in the evolving stage. Agricultural farmers worldwide commonly utilize the process of cultivating and harvesting crops to produce food and fiber. Therefore, crop prediction is vital for smart agriculture and the proposed approach involves utilizing all the necessary resources to facilitate crop growth and maintenance. Crop cultivation used to be carried out based on farmers' actual experience. For farmers and agricultural decision-makers to make prompt and accurate judgments that will impact the caliber of agricultural harvests, advanced tools are required. A prediction has been supplanted by machine learning techniques as a result. To anticipate agricultural yield, the authors provide a prediction system based on an ensemble of machine learning classifiers. Also, they discovered that an ensemble technique provided more accurate prediction than using the currently available categorization algorithms.

1 INTRODUCTION

Most nations' economies rely on the agriculture industry, which is vital for ensuring food security and livelihood. Despite its importance, the industry faces several

DOI: 10.4018/978-1-6684-9975-7.ch007

obstacles such as climate change, pests, diseases, and erratic weather patterns that adversely impact crop yields and food production (Fenu & Malloci, 2019), (Pant et al., 2021). Precise and dependable crop prediction is critical in addressing these challenges and safeguarding food security. Predicting crops involves evaluating the projected yield for each season by considering factors such as soil type and weather conditions (Kalimuthu et al., 2020).

Historically, predicting crop yields has been accomplished through manual means, which are slow, ineffective, and vulnerable to mistakes. However, due to the availability of substantial data and technological advancements, machine learning (ML) has become a promising alternative for precise and efficient crop prediction (Van Klompenburg et al., 2020). ML algorithms can examine vast amounts of information, recognize patterns, and provide precise predictions (Nischitha et al., 2020). Numerous investigations have been conducted to design ML-based models for crop prediction, utilizing various techniques such as neural networks, decision trees, support vector machines, and Bayesian networks (Gupta et al., 2021).

Ensemble techniques have become increasingly popular for enhancing the precision and stability of ML models in recent times (Kumar et al., 2015), (Rajak et al., 2017). For more accurate predictions than single models, ensemble approaches combine multiple models. Ensemble techniques can compensate for the shortcomings of individual models and deliver more durable and dependable predictions by using multiple models (Dharani et al., 2021).

The primary focus of this study is crop prediction for smart agriculture using an ensemble technique that combines best-performing ML algorithms such as Decision Tree, Logistic Regression, Random Forest, Gaussian Naïve Bayes, and K-Nearest Neighbour. These algorithms were chosen based on their popularity, simplicity, and effectiveness in classification tasks (Reddy & Kumar, 2021). The objective of this study is to develop a dependable and precise ML-based model for crop prediction utilizing an ensemble technique. The model is trained and assessed on a dataset comprising past crop yields and weather data for a particular region. The model's precision is evaluated utilizing metrics like accuracy, ROC, and F1 score. The outcomes of this study are compared to other research studies in crop prediction using ML.

This chapter addresses the significant concern of dependable and accurate crop prediction by utilizing an ensemble technique that merges five ML algorithms. The research makes a valuable contribution to the area of agriculture and ML by presenting an effective and precise strategy for crop prediction that has the potential to significantly improve crop yields and food production. The results of this study have the potential to guide policymakers, farmers, and stakeholders in making informed decisions and implementing strategies to boost food security and combat the effects of climate change on the agriculture industry.

The remaining sections of this chapter are structured as follows: In Section 2, we present an overview of the existing work in crop prediction utilizing ML. Section 3 outlines the methodology employed in this study, including the dataset, preprocessing, and ensemble technique. In Section 4, we present the findings of the study, including the performance of the individual models and the ensemble model. Finally, in Section 5, the outcomes and concluding remarks of the study are presented.

1.1 Related Works

This section discusses the related work pertaining to the current study. Vazquez-Carillo et al. proposed a ML approach for predicting crop yields using satellite imagery (Vazquez-Carillo., 2020). The authors used a random forest model to predict yields for three different crops: wheat, soybeans, and corn. In this study, the authors used satellite imagery from the European Space Agency's Sentinel-2 satellite, which captures images of the Earth's surface at a spatial resolution of 10 meters. They used data from the growing season of each crop, which is defined as the period from planting to harvest. The satellite imagery was processed to extract various features related to vegetation indices, land cover, and climate variables. The authors trained a random forest model on this data to predict the yield for each crop. They compared their results to those of traditional statistical models commonly used for crop yield prediction, such as multiple linear regression and principal component regression (Kim et al., 2014). The experimental results exhibited that the random forest model outperformed these conventional models, achieving a mean absolute error of 0.14, 0.11, and 0.18 for wheat, soybeans, and corn, respectively. The study also analyzed the importance of different features in predicting crop yields. The authors found that vegetation indices, such as the normalized difference vegetation index (NDVI), were the most important features for predicting yields. The study showed that the use of satellite imagery and machine learning can provide accurate and timely predictions of crop yields, which can help farmers and policymakers make more informed decisions related to crop management and food security. Overall, this study demonstrates the potential of machine learning and satellite imagery for crop yield prediction. By using such techniques, farmers and policymakers can gain vital knowledge on agricultural productivity and be better equipped to manage crops and ensure food security.

Zhang et al. in their work, provided an overview of recent studies on crop yield prediction using machine learning (Zhang et al., 2020). The authors analyzed various machine learning algorithms used in crop yield prediction, including decision trees, random forests, support vector machines, artificial neural networks, and deep learning (Agarwal & Tarar, 2021). They investigated the advantages and limitations of each approach and highlighted the need for more research to identify the most effective algorithms for different crops and regions (Champaneri et al., 2016). The article also considered the data sources used in crop yield prediction, such as weather data, soil data, and satellite imagery (Kumar et al., 2020). The authors emphasized the importance of high-quality data and data integration in achieving accurate yield predictions. Additionally, the article discussed the performance metrics used in crop yield recommendation, such as mean absolute error, root means squared error, and coefficient of determination (Abbas, Afzaal, Farooque, & Tang, 2020). The authors highlighted the need for standardized evaluation metrics to compare the performance of different models.

The review article also discussed the applications of crop yield prediction using machine learning, such as precision agriculture, crop insurance, and food security (Palanivel & Surianarayanan, 2019). The authors emphasized the potential of machine learning for improving agricultural practices and increasing crop yields (Ghadge et al., 2018). Overall, the article provides a comprehensive overview of recent studies on crop yield prediction using machine learning (Muruganantham et al., 2022). The article highlights the potential of machine learning for improving crop yield prediction and the need for more research to identify the most effective approaches for different crops and regions.

Singh et al. in their work presented a detailed and comprehensive review of recent research on crop yield prediction using machine learning techniques (Dharani et al., 2021). The authors begin by highlighting the significance of crop yield prediction in agriculture and the potential benefits of using machine learning for this purpose. They then provide an overview of the different types of machine learning algorithms commonly used for crop yield prediction, including supervised and unsupervised learning methods (Rao et al., 2022). The authors also discussed the different types of data sources used in crop yield prediction research, such as climate data, soil data, and remote sensing data, and how these data sources can be integrated to improve prediction accuracy. They further explore the evaluation metrics used in crop yield prediction studies and provide a critical analysis of their strengths and weaknesses. The authors highlighted the importance of developing standardized datasets and the need for more transparency and interpretability in machine learning models to facilitate their adoption in practical applications. They also discuss the challenges associated with collecting high quality data in agricultural settings and suggest potential solutions, such as the use of crowdsourcing and citizen science approaches (Abbas, Afzaal, Farooque, & Tang, 2020). For scholars and practitioners interested in crop yield prediction using machine learning approaches, this review paper is a helpful resource. It highlighted the potential benefits and challenges associated with using ML in smart agriculture and identified the important areas for future research and development. Compared to the above works the proposed work is different in the classification task which combines the best-performing machine learning classifiers to predict the type of the crop.

1.2 Proposed Methodology

Figure 1. Block diagram of the proposed crop prediction system

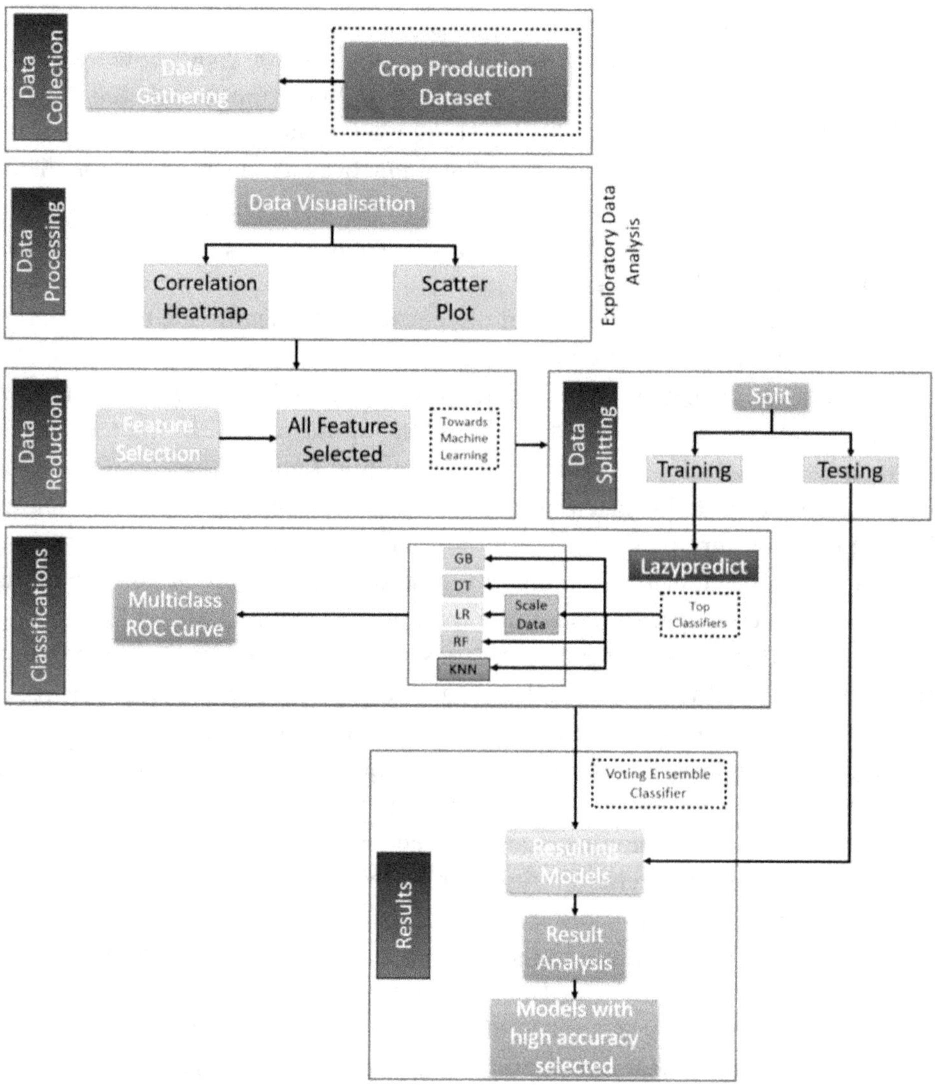

1.2.1 Data Collection

The dataset is constructed by taking Crop Production Dataset which is taken from the well-known data repository, Kaggle. This dataset was created by enhancing rainfall, climate, and fertilizer data sets that were previously accessible to India. With over 2000 scenarios collected as a text file along with 7 special features (N, P, K, temperature, humidity, pH, and rainfall) and the final column 'label' selected as the output, with 22 classifications of crops ('rice', 'maize', 'chickpea', 'kidneybeans', 'pigeonpeas', 'mothbeans', 'mungbean', 'blackgram', 'lentil', 'pomegranate', 'banana', 'mango', 'grapes', 'watermelon', 'muskmelon', 'apple', 'orange', 'papaya', 'coconut', 'cotton', 'jute' and 'coffee').

1.2.2 Data Processing

The data input into the computer undergoes some machine learning algorithms to make it more usable i.e., making it more meaningful and informative. We use two such algorithms, namely Correlation Heatmap and Scatter Plot. There is a correlation between two variables when they are closely related to each other. The two primary types of correlations are positive and negative correlations. In correlation, when one variable increases, the other increases as well; when one changes direction, the other changes as well. A negative correlation occurs when two variables change in opposing directions, for instance, when one increases while the other decreases. Depending on the correlation coefficient, the value can range from -1 to 1. The correlation coefficient is computed as follows:

$$Corr_{Co} = \frac{\Sigma\left(px_i - \overline{px}\right)\left(py_i - \overline{py}\right)}{\sqrt{\Sigma\left(px_i - \overline{px}\right)^2 \Sigma\left(py_i - \overline{py}\right)^2}} \tag{1}$$

The correlation formula allows for the calculation of the correlation coefficient between two variables, px and py. The mean values of px and py are represented by $\overline{px}$ and $\overline{py}$, respectively. Different values of px and py are represented by px_i and py_i.

Scatter plots depict the correlation between two variables or the extent to which one variable impacts the other, using dots in a two-dimensional plane. They share similarities with line graphs as they also display data points along the horizontal and vertical axes. In this graph, we set the 'N' feature as our x-axis and 'P' as the y-axis and we see the N and P values at which our different crops grow.

1.2.3 Data Reduction

Sometimes the collected data will have null or empty values which need to be cleaned and we need to delete redundant features by using SelectKBest by which we can make the prediction using a minimum number of features before we can move ahead with the evaluation of our values. Upon checking our dataset, we find that it had no null value; and using SelectKBest we see that all 7 features are useful thus we need to use them all. SelectKBest works by retaining the first k features of X with the highest scores.

1.2.4 Data Splitting

The data is divided into two subsets, mainly the training and the testing. Training is used to train the model and testing is used to evaluate the data. Splitting ensures that the data models created using the training data are accurate which is checked with the help of the testing data. We split the data in a ratio of 7:3 (training: testing). We use random_state to ensure the output after splitting the data is the same every time.

1.2.5 Classification Techniques

The library "lazypredict" is utilized to simplify the process of building basic models with minimal code, and to assist in identifying the most effective models without requiring parameter tuning. So, using it we found the top 5 classification models based on which we develop the rest of the model.

1.2.5.1 Gaussian Naïve Bayes

The Bayes theorem-based probabilistic machine learning method known as naive Bayes is commonly used in a variety of classification applications. In real-world circumstances, there may be several X variables, hence the Bayes rule provides a formula for determining the probability of Y given the condition X, which is calculated as follows:

$$P\left(X|Y=c\right)=\frac{1}{\sqrt{2\pi\sigma_c^2}}e^{\frac{-(x-\mu_c)^2}{2\sigma_c^2}} \tag{2}$$

Naive Bayes can be applied when the features are independent of one another, meaning that each X variable stands alone. This technique is more effective than the Bayes formula.

When X is a continuous variable, we can use the probability density function of a distribution to determine the likelihood, if X follows a specific distribution. If X has a Gaussian or normal distribution, the probability density function can be replaced with the normal distribution's probability density, referred to as Gaussian Naive Bayes. To calculate this formula, the mean and variance of X for a given class c of Y must be determined, denoted by sigma and μ, respectively. By replacing the parameters with a new input value for the variable, the Gaussian probability density function can be utilized to create predictions. Thus, the Gaussian function estimates the likelihood of the new input value.

1.2.5.2 Decision Tree

The decision tree is a well-liked and effective method for categorizing and predicting data. Each internal node denotes a test on a feature, each branch denotes a test result, and each terminal node also called the leaf node denotes a class label. It adopts a tree-like form that mimics a flowchart.

To classify an instance using the decision tree, the procedure begins at the root node of the tree, evaluating the attribute specified by the node, and moving down the tree's branch that matches the attribute's value. Then, the subtree rooted at the new node undergoes the same process again. The choice of attributes is the main difficulty in creating a DT. That is, it is challenging to choose which qualities to employ as root nodes or internal nodes given the abundance of available attributes. Information Gain and Gini Index are two methods that can be used to this end:

$$InfoGain(T,X) = Entropy(T) - Entropy(T,X) \tag{3}$$

where T refers to the current state and X to the selected attribute;

$$GiniIndex = 1 - \Sigma(p) = 1 - \left[(p+)^2 + (p-)^2\right] \tag{4}$$

where p+ represents the probability of Yes/Good and p- the probability of No/Bad.

1.2.5.3 Random Forest

In order to improvise the model's classification accuracy, the Random Forest classifier uses numerous decision trees that have been trained on different subsets of the input data. Instead of relying just on one decision tree, the Random Forest uses forecasts from all of the trees and bases its prediction on the predictions that received the most votes.

Because Random Forest is straightforward to grasp, it is very well-liked. Classification and regression can also be carried out with it. In addition, the model's accuracy is greatly improved, and the overfitting problem is solved, even for huge, highly dimensional data sets. Data extrapolation cannot be done using Random Forest since the findings could be unreliable. Although Random Forest may be used for both classification and regression problems, classification jobs are where it shines. Moreover, it fails to deliver accurate findings when working with sparse data and it takes more resources, larger data, and more time for implementation.

1.2.5.4 K-Nearest Neighbour (KNN)

In supervised learning, the K-nearest neighbor algorithm is a method for making predictions about the grouping of data points based on proximity. Due to its assumption that similar points can be found near one another, it is typically used as a classification algorithm. It is important to note that the KNN technique only stores a training dataset and does not go through a training phase, placing it in the category of "lazy learning" models. Additionally, this suggests that all calculations are completed at the same time as a prediction or categorization. It is also known as an instance-based or memory-based learning strategy because it significantly relies on memory to maintain all of its training data.

To identify the data points that are nearest to a particular query point, it is necessary to compute the dissimilarity between the query point and the other data points. The commonly used distance metric is Euclidean distance, which applies only to real-valued vectors. The straight-line distance using the Euclidean distance is determined between the query location and the point being measured.

K-NN techniques determine classification by examining neighboring points to determine the number of neighbors that should be examined. The closest neighbor instance will be assigned to the same class if $k = 1$. To prevent overfitting or underfitting, it is necessary to consider multiple values of k when setting it. The higher the value of k, the lower the variance, while the smaller the value of k, the higher the variance, but the lower the bias. High values of k may be better for data with more outliers or noise, thus influencing the choice of k. The optimal value for k is generally determined by cross-validation techniques and odd values of k.

1.2.5.5 Logistic Regression

The initial step involves scaling the dataset utilizing the StandardScaler, which performs standardization. Standardisation helps to ensure that the individual features are roughly normally distributed with a mean of zero and a variance of one, which is a typical requirement for many machine learning estimators. The output is predicted using logistic regression in the case of a categorical dependent variable.

It is essential that the output in this case be discrete or categorical. In contrast to the precise values of 0 and 1, yes or no, true or false, etc., it provides probabilistic values between 0 and 1. When calculating the probability in logistic regression, the logistic or sigmoid function is used. The logistic function is a straightforward S-shaped curve that converts input into a value between 0 and 1.

1.3 Result and Discussions

This section discusses the experimental results obtained using the proposed model. Firstly, the relationship between the features is visualized using the correlation

Figure 2. Relationship between features

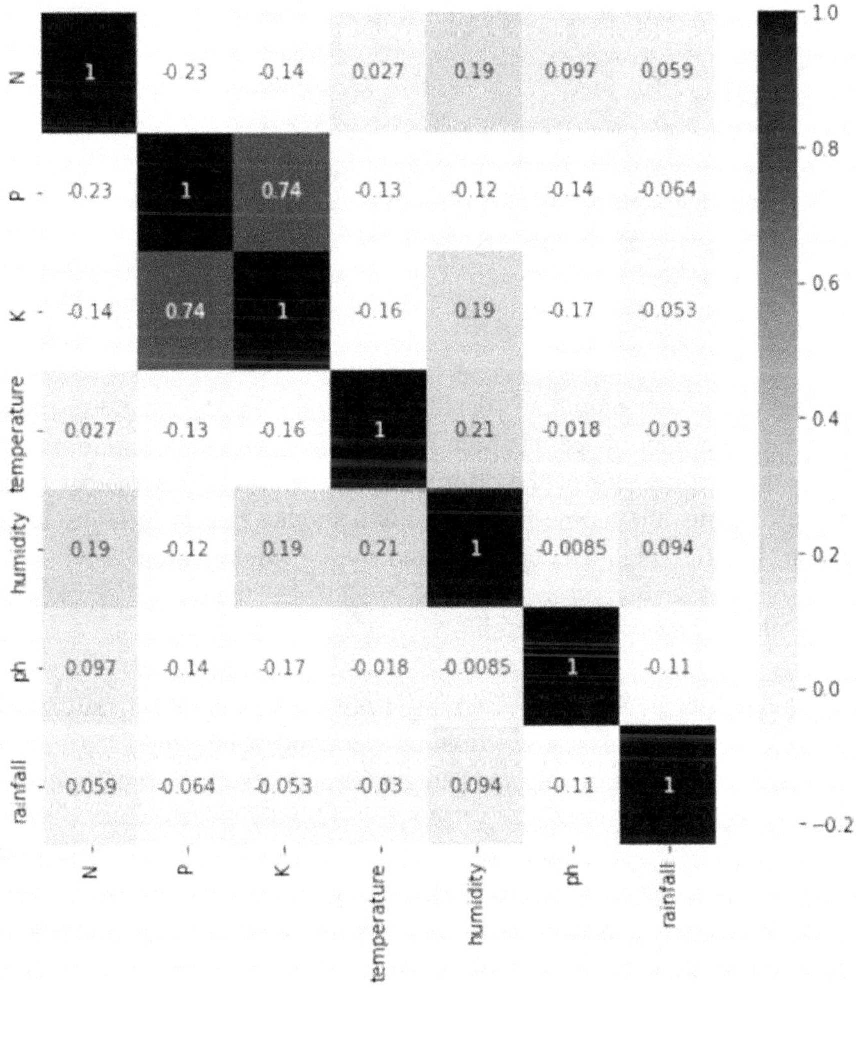

Figure 3. Relationship between features

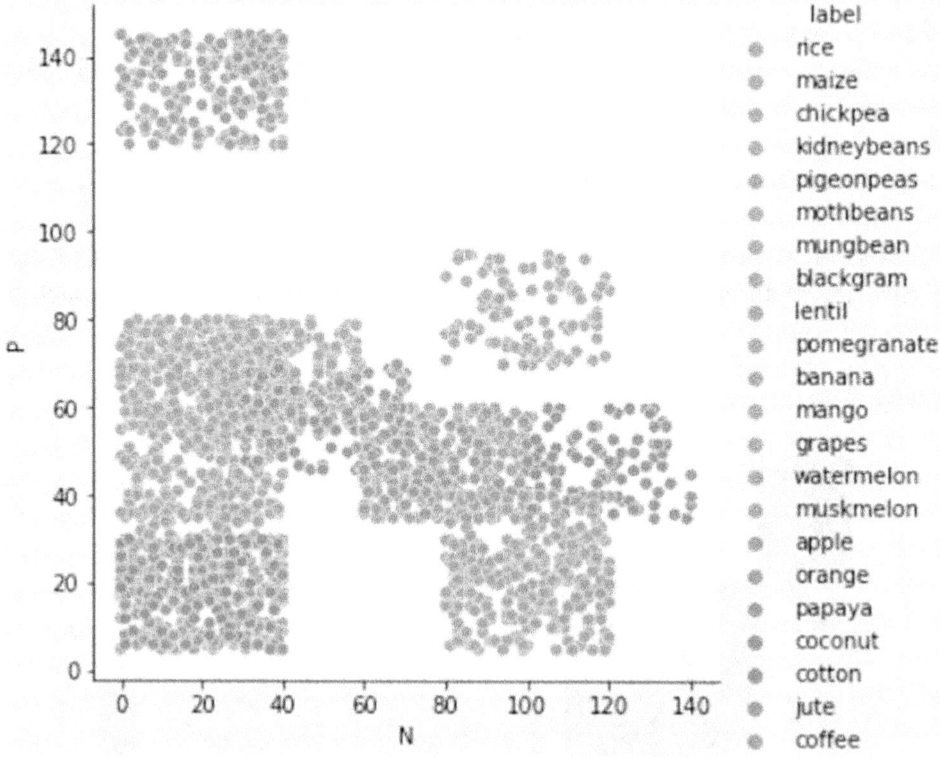

coefficient and scatter plot that is depicted in Figure 2 and Figure 3 respectively. Then the feature importance has been plotted and depicted in Figure 4. Table 1 tabulates the performance of the classifiers after executing lazypredict function. From these results, the top-performing classifiers have been selected to build the ensemble classifier. Figure 5 depicts the top five performing classifiers.

1.3.1 Voting Ensemble Classifier

A voting classifier is a machine learning estimator that combines multiple base models or estimators to make predictions by averaging their results. The predictions of each estimator can be aggregated using different voting criteria. There are two types of voting criteria:

- Hard voting, also known as majority voting, is a simple technique where the final prediction is made by selecting the most common prediction among the individual models. For example, if there are three models and two of them

Figure 4. Feature importance

```
Importance of N is 634.109614
Importance of P is 1370.832913
Importance of K is 19278.607456
Importance of temperature is 65.061410
Importance of humidity is 2255.437627
Importance of ph is 47.624288
Importance of rainfall is 478.509208
```

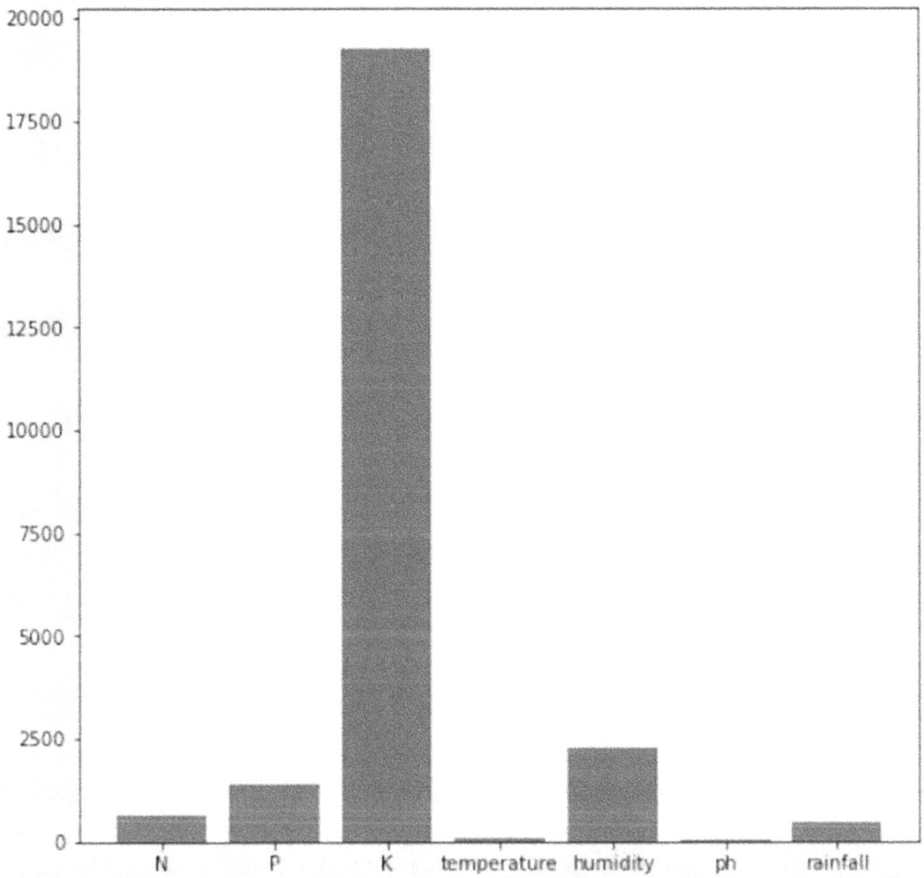

predict the class "A," while the third predicts the class "B," the hard voting approach would choose a class "A" as the final prediction. Hard voting is most effective when the individual models are diverse and make independent errors. In other words, if the models are making the same mistakes, hard voting may not improve the overall performance.

Figure 5. Top five performing classifiers

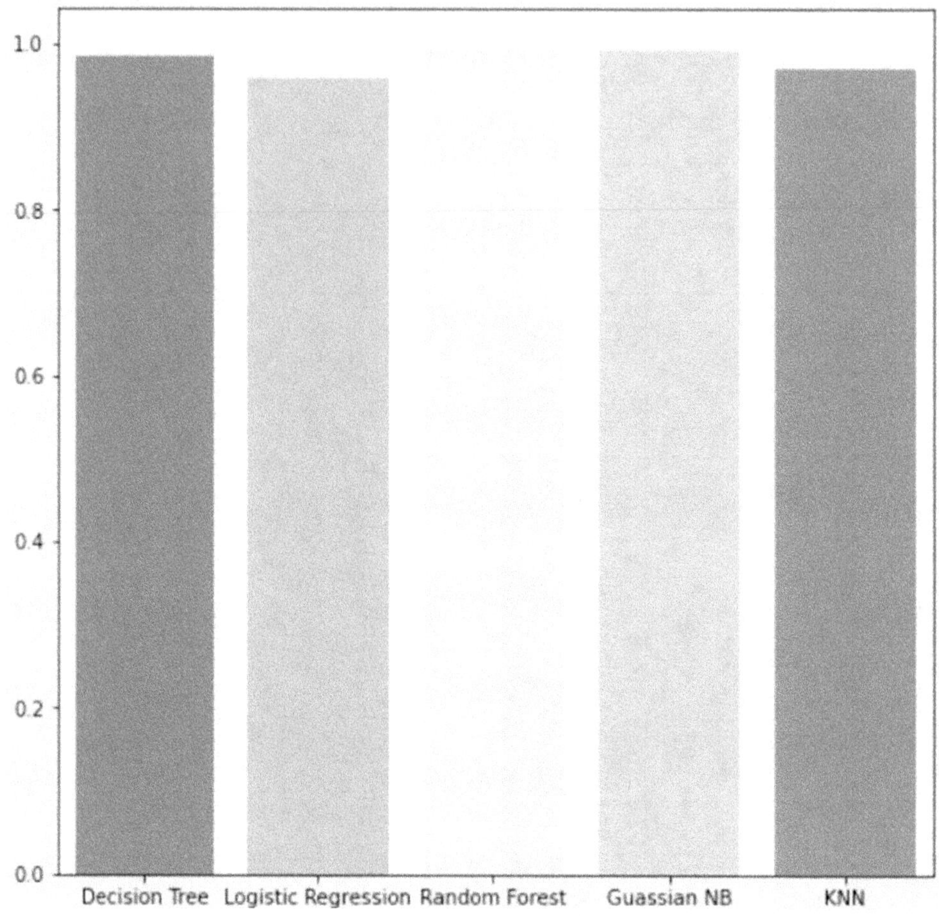

- Soft voting, on the other hand, is a more sophisticated technique where the final prediction is based on the probability estimates of each model. Each model assigns a probability to each class, and the final prediction is made by taking the average of these probabilities across all models. Soft voting is more robust than hard voting when individual models have high accuracy but low confidence. By considering the probability estimates, soft voting can make more informed decisions and provide more accurate predictions.

We made four voting ensemble classifiers:

Table 1. Performance of classifiers using LazyPredict

Model	Accuracy	Balanced Accuracy	F1 Score	Time Taken
GaussianNB	0.99	0.99	0.99	0.02
RandomForestClassifier	0.99	0.99	0.99	0.39
QuadraticDiscriminantAnalysis	0.99	0.99	0.99	0.03
DecisionTreeClassifier	0.99	0.99	0.99	0.03
BaggingClassifier	0.99	0.99	0.99	0.08
ExtraTreesClassifier	0.98	0.99	0.98	0.20
LGBMClassifier	0.98	0.98	0.98	1.23
LabelPropagation	0.97	0.97	0.97	0.11
LabelSpreading	0.97	0.97	0.97	0.18
SVC	0.97	0.97	0.97	0.19
LinearSVC	0.96	0.96	0.96	0.11
CalibratedClassifierCV	0.96	0.96	0.96	0.54
KNeighborsClassifier	0.96	0.96	0.96	0.03
LogisticRegression	0.96	0.96	0.96	0.20
LinearDiscriminantAnalysis	0.95	0.95	0.95	0.03
NuSVC	0.94	0.94	0.94	0.36
SGDClassifier	0.88	0.89	0.89	0.09
ExtraTreeClassifier	0.88	0.88	0.88	0.02
NearestCentroid	0.86	0.86	0.85	0.02
Perceptron	0.86	0.86	0.86	0.05
PassiveAggressiveClassifier	0.81	0.81	0.81	0.05
BernoulliNB	0.78	0.78	0.77	0.02
RidgeClassifier	0.64	0.66	0.57	0.02
RidgeClassifierCV	0.64	0.66	0.56	0.04
AdaBoostClassifier	0.21	0.23	0.17	0.21
DummyClassifier	0.03	0.05	0.00	0.02

1. In the first model, we combined Decision Tree, Logistic Regression, Random Forest, Gaussian Naïve Bayes, and K-Nearest Neighbour classification using the soft voting method and achieved an accuracy of 100%.
2. In the second model, we combined Decision Tree, Gaussian Naïve Byes, and Random Forest classification using the soft voting method and achieved an accuracy of 100%.

3. In the third model, we combined Decision Tree, Logistic Regression, Random Forest, Gaussian Naïve Bayes, and K-Nearest Neighbour classification using the hard voting method and achieved an accuracy of 70.7%.
4. In the fourth model, we combined Decision Tree, Gaussian Naïve Byes, and Random Forest classification using the hard voting method and achieved an accuracy of 100%.

1.3.2 Multiclass ROC Curve

ROC curves show the effectiveness of binary classification models graphically. A one-vs-all method can be used to assess the effectiveness of a multiclass classification

Figure 6. ROC plots of proposed approach

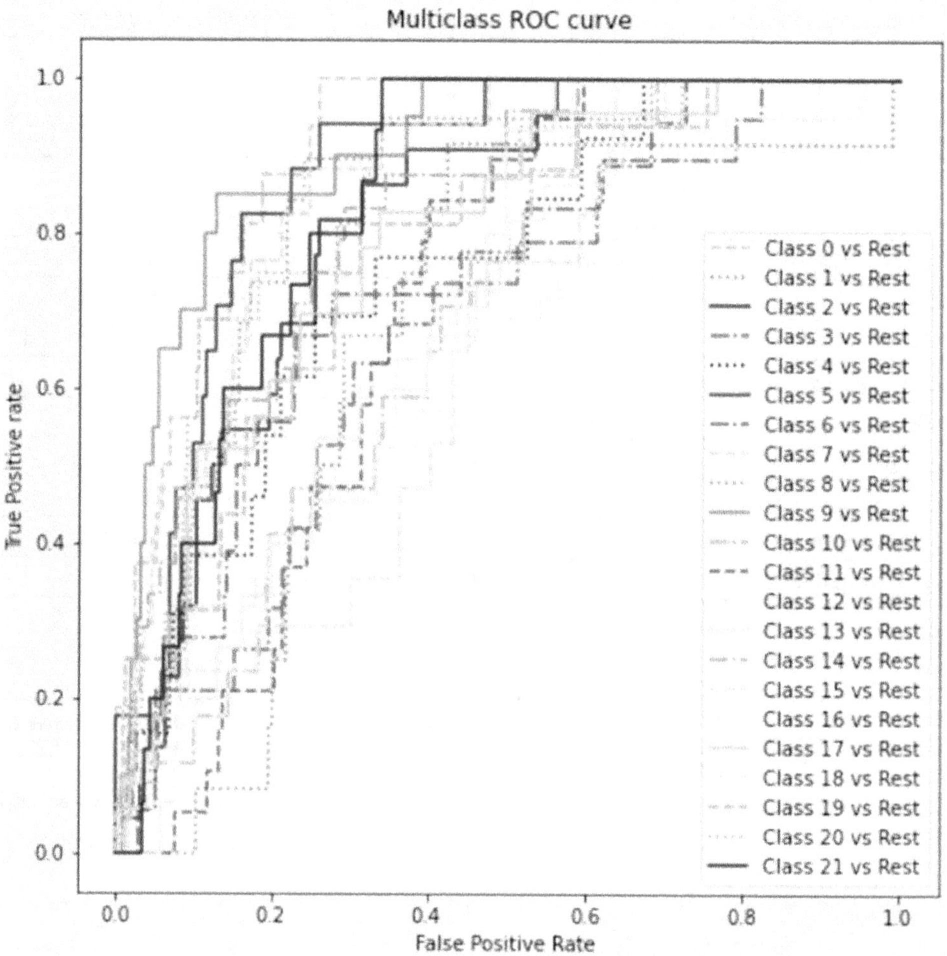

model by extending the ROC curve. Figure 6 depicts the ROC plots of the proposed approach. In a multiclass classification problem, there are multiple classes, and each class can be treated as a binary classification problem, where one class is positive, and the rest of the classes are negative classes. In order to generate the ROC curve, the true positive rate (sensitivity) is plotted along with the false positive rate (1-specificity) for each class. The multiclass ROC curve can be used to compare the performance of different multiclass classification models, by selecting an appropriate threshold for each class, and identifying which classes are harder to predict.

2. CONCLUSION

In conclusion, the chapter presented an efficient and accurate approach for crop prediction using machine learning techniques. By utilizing ensemble techniques, two models were developed which were able to predict the correct crop with 100% accuracy. This approach has the potential to significantly improve the crop yield and economic output of the agriculture industry. Additionally, the research demonstrates the power and potential of machine learning in agriculture, particularly in crop prediction. Further research can be conducted to expand the dataset and explore the use of other machine-learning techniques for crop prediction. Overall, this research paper contributes to the field of agriculture and machine learning and provides valuable insights for future developments in this area.

REFERENCES

Abbas, F., Afzaal, H., Farooque, A. A., & Tang, S. (2020). Crop yield prediction through proximal sensing and machine learning algorithms. *Agronomy (Basel)*, *10*(7), 1046. doi:10.3390/agronomy10071046

Agarwal, S., & Tarar, S. (2021). A hybrid approach for crop yield prediction using machine learning and deep learning algorithms. *Journal of Physics: Conference Series*, *1714*(1), 012012. doi:10.1088/1742-6596/1714/1/012012

Champaneri, M., Chachpara, D., Chandvidkar, C., & Rathod, M. (2016). Crop yield prediction using machine learning. *Technology*, *9*, 38.

Kaggle. (2023). *Crop Production Dataset*. Kaggle. https://www.kaggle.com/datasets/atharvaingle/crop-recommendation-dataset

Dharani, M. K., Thamilselvan, R., Natesan, P., Kalaivaani, P. C. D., & Santhoshkumar, S. (2021, February). Review on crop prediction using deep learning techniques. [). IOP Publishing.]. *Journal of Physics: Conference Series, 1767*(1), 012026. doi:10.1088/1742-6596/1767/1/012026

Fenu, G., & Malloci, F. M. (2019, November). An application of machine learning technique in forecasting crop disease. In *Proceedings of the 3rd International Conference on Big Data Research* (pp. 76-82). 10.1145/3372454.3372474

Ghadge, R., Kulkarni, J., More, P., Nene, S., & Priya, R. L. (2018). Prediction of crop yield using machine learning. *Int. Res. J. Eng. Technol., 5.*

Gupta, A., Nagda, D., Nikhare, P., & Sandbhor, A. (2021). Smart crop prediction using IoT and machine learning. *International Journal of Engineering Research & Technology (Ahmedabad)*, 2278–0181.

Kalimuthu, M., Vaishnavi, P., & Kishore, M. (2020, August). Crop prediction using machine learning. In *2020 third international conference on smart systems and inventive technology (ICSSIT)* (pp. 926-932). IEEE. 10.1109/ICSSIT48917.2020.9214190

Kim, Y. H., Yoo, S. J., Gu, Y. H., Lim, J. H., Han, D., & Baik, S. W. (2014). Crop pests prediction method using regression and machine learning technology: Survey. *IERI Procedia, 6*, 52–56. doi:10.1016/j.ieri.2014.03.009

Kumar, R., Singh, M. P., Kumar, P., & Singh, J. P. (2015, May). Crop Selection Method to maximize crop yield rate using machine learning technique. In 2015 international conference on smart technologies and management for computing, communication, controls, energy and materials (ICSTM) (pp. 138-145). IEEE. doi:10.1109/ICSTM.2015.7225403

Kumar, Y. J. N., Spandana, V., Vaishnavi, V. S., Neha, K., & Devi, V. G. R. R. (2020, June). Supervised machine learning approach for crop yield prediction in agriculture sector. In *2020 5th International Conference on Communication and Electronics Systems (ICCES)* (pp. 736-741). IEEE. 10.1109/ICCES48766.2020.9137868

Muruganantham, P., Wibowo, S., Grandhi, S., Samrat, N. H., & Islam, N. (2022). A systematic literature review on crop yield prediction with deep learning and remote sensing. *Remote Sensing (Basel), 14*(9), 1990. doi:10.3390/rs14091990

Nischitha, K., Vishwakarma, D., Ashwini, M. N., & Manjuraju, M. R. (2020). Crop prediction using machine learning approaches. *International Journal of Engineering Research & Technology (Ahmedabad), 9*(08), 23–26.

Palanivel, K., & Surianarayanan, C. (2019). An approach for prediction of crop yield using machine learning and big data techniques. *International Journal of Computer Engineering and Technology*, *10*(3), 110–118. doi:10.34218/IJCET.10.3.2019.013

Pant, J., Pant, R. P., Singh, M. K., Singh, D. P., & Pant, H. (2021). Analysis of agricultural crop yield prediction using statistical techniques of machine learning. *Materials Today: Proceedings*, *46*, 10922–10926. doi:10.1016/j.matpr.2021.01.948

Rajak, R. K., Pawar, A., Pendke, M., Shinde, P., Rathod, S., & Devare, A. (2017). Crop recommendation system to maximize crop yield using machine learning technique. *International Research Journal of Engineering and Technology*, *4*(12), 950–953.

Rao, M. S., Singh, A., Reddy, N. S., & Acharya, D. U. (2022). Crop prediction using machine learning. *Journal of Physics: Conference Series*, *2161*(1), 012033. doi:10.1088/1742-6596/234/1/012033

Reddy, D. J., & Kumar, M. R. (2021, May). Crop yield prediction using machine learning algorithm. In *2021 5th International Conference on Intelligent Computing and Control Systems (ICICCS)* (pp. 1466-1470). IEEE. 10.1109/ICICCS51141.2021.9432236

Van Klompenburg, T., Kassahun, A., & Catal, C. (2020). Crop yield prediction using machine learning: A systematic literature review. *Computers and Electronics in Agriculture*, *177*, 105709. doi:10.1016/j.compag.2020.105709

Vazquez-Carillo. (2020) *Crop Yield Prediction using Machine Learning and Satellite Imagery*.

Zhang, C., Marzougui, A., & Sankaran, S. (2020). High-resolution satellite imagery applications in crop phenotyping: An overview. *Computers and Electronics in Agriculture*, *175*, 105584. doi:10.1016/j.compag.2020.105584

Chapter 8
An Integrated Approach for Selection and Design of Sustainable Farmers' Protective-Hat in India

Suchismita Satapathy
https://orcid.org/0000-0002-4805-1793
KIIT University, India

ABSTRACT

Normally, the farmers in India are required to work in adverse climatic conditions while performing their agricultural activities. Though a number of preventive measures are available for the protection of farmers, the nominal and small scale farmers are still unable to bear these expenses. Therefore, an attempt was made in this study to the select the most appropriate material using Fuzzy-TOPSIS method to carry out the design of the head-cap for farmers in terms of parameters like stiffness, cost, durability, and availability in India. Further, for a sustainable design of these head-cap, the genetic algorithm (GA) as well as Sun Flower Optimization (SFO) method were use to obtain the best fitness values based on the anthropometric dimensions of farmers' head.

1. INTRODUCTION

A variety of hat exists for wearing in different contexts like bicycle helmets, surgical caps, fedoras, yarmulkes, and ushankas, etc. and has different inherent functions apart from providing protection. Thus an exciting opportunity is provided by this

DOI: 10.4018/978-1-6684-9975-7.ch008

diversity of hats to provide the form of design in addition to hat technology creation. This paper discusses the evaluation of hat technology through proper and suitable material selection. Different literature exists regarding the protective equipment in agriculture. For instance, (Meeker et al., 2002) have explored the preventive measures as well as consistency in using protective equipment by farm women with varying levels of time commitment in farm activities in south-east Louisiana. It was found that women were less likely to wear personal protective equipment such as wearing of seat belts on tractors, wearing of boots or gloves, and hearing or breathing protection. However, no difference was found between groups for use of protective equipment such as hair holders, protection of skin, or helmets on all terrain-vehicles.

According to, Safety and health in agriculture (2011) the supplementary protection are provided by personal protective equipment in agricultural production against exposure to hazardous conditions where by other means the safety of workers cannot be ensured. For the open air workers, the bamboo as well as leaf headgear are the most indispensable items that are produced throughout the Assam State in India Saikia, (2012). (Makalew et al., 2013) have revealed that the farmer characteristics such as education level, age, family member numbers, distance with organic manure and getting in touch with instructor influenced the motivation of farmers. Throughout the world hats exist in nearly all cultures which reflect rich traditions as well as self extension over generations (Chico, 2013).

(Jazi et al., 2014) have made a computational learn of a helmeted humans head, by studying the effect of helmet pad material on the stage of accelerations, shear stresses, and inflicted pressures in a human head model subjected to a ballistic-impacts. It was found that a pad with lesser stiffness was more effective to absorb the impact energy, reduced sudden accelerations of the head, and subsequently lowers the level of brain injuries. The producers, consumers, and policymakers interests as well as the widespread belief in the effects of health benefits of organic agriculture encourages a greater resource allocations in modern society for carrying out research in this area (Aleixandre et al., 2015). (Jallow et al., 2017) have reported of 58% farmers not using of any personal protective equipment while dealing with pesticides.

Barman et al., (2017) have introduced the invent of a low cost "agro-quadcopter" module in order to receive command & send feedbacks and to react spontaneously to administer through a range of control system when working in real-time. Further it was revealed to be able of performing the study on each day -weather, each day temperature records, detection of infected crops, keeping pests away, aiding in artificial pollination, as well as spraying & seeding, thereby reducing farmers' workload and enhancing the farmers' productivity.

Dutta (2018) has discussed about losing threats of traditional knowledge systems in Assam region of North-East India due to proliferation of neo-societal norms, human migration, because of social acceptance changing norms of art-craft with imitation products imported from mass manufacturing facilities. Wästerlund (2018) has discussed about the implication of heat stresses in agricultural working-environments affecting the human body, the posing risks to human health, its measurement, affects on labour productivity, as well as its management. (Khan et al., 2019) have evaluated the materials such as "Acrylonitrile Butadiene Styrene", "Epoxy glass fibre", "Polycarbonate", and "Polycarbonate with 30% glass" for helmet shell by the use of "Finite Element Analysis". Considering the Assamese culture & society, the Japi plays a pivotal role, which is a beautiful constructional realization by man. While working in the farm lands, in order to protect the farmers from the rain and heat, the invention of Japi was done (Nath, 2019).

With the development of wearable technologies growing, a number of devices are found situated on our head, but many challenge hinder from extensive adoptions because of their embarrassed, bulky form factors, socially stigmatized invent, and a require of a well-developed head-based contact design language as well. (Dierk et al., 2019) have explored a large, socially acceptable, and head-worn interactive wearable hat based on a gesture elicitation study of seventeen participants. The decision makers can evaluate the existing or potential alternatives through MCDM uses by applying various conflicting criterion simultaneously (Belton and Stewart, 2002; Kou et al., 2011; Zhou et al., 2010). Moreover, Zulqarnain and Dayan (2017) have proposed Fuzzy-TOPSIS method for decision making. Thus, an effort has been made in this study to make a selection of most appropriate available material in terms of the parameters like stiffness, cost, durability, and availability, in India to design the head cap for the farmers.

As the farmers as well as agricultural-workers are unique among the occupational groups living and working exclusively in rural-areas, therefore understanding of the nature of rural-areas in view of health and health-care become very important (Gregoire, 2002). In India there has been an ever growing suicidal case of farmers. The increase in suicide-rate of some particular group in society often indicates of the general environmental factors that place selective-pressures on the group. These factors vary in type from the demographic (family-structure), to the psychological (e.g. mental-illness) to the economic (e.g. nonstop crop-failures and debt) (Cavanagh et al., 2003). In the rural-areas including the farming community, the mental-health problems and services are almost neglected. Nearly one-sixth of all health related disorders are constituted owing to mental illness (Das, 2011).

There have been a number of studies that focus on economic as well as political aspects for suicidal cases of farmers in India, while taking in to consideration of the mental-health aspects of farmers are very few. Therefore, an attempt was made in the present study to identify the stressors for mental-illness among farmers that lead to their loosing of self-control as well as depression levels. Normally a person faces many kind of stress in day to day life, may be due to work place, personal life or in social relation. Budgeting, family life, keeping up with local trends and future planning are all challenges that arise on a daily basis Social isolations and working for long hours can generate stress.

Stress is nothing but an emotional or physical reaction. It changes mood of a person and behaviour also. It can be generated from any kind of action or thinking that hurts sentiment of a person and put him in uncomfortable situation. It raises anger of a person, makes him tense and upset. It is the response to a requirement or any kind of provocation. Stress has also many negative impacts like mental instability, unable to take pressure in work place, unable to focus and increase of physical sickness etc. The results of mental stress in work place are reduction of job satisfaction, reduction of performance and work interest also. Stress can appear from a requirement, chance, restriction, pressure or challenges. Stress is harmful at individual level and also in group. It may be the initiator of any fight or any bad habits.Stress is not an illness, however when enduring and with a specific force, it might likewise prompt a procedure of mental or physical illness, for example, cardiovascular infections, diabetes, fit of anxiety, gloom, musculoskeletal complaints, brought down invulnerability, and burnouts. They may feel like powerless and sad and chose for self-destructive plans. Normally people experience stress related to the environment, psychological states, and social issues (Padhy & Raju, (2018)). A Stressed individual are always under constant threat, it may be the cause of medical complications like the imbalance pulse-rate, obesity, diabetes, and imbalance blood-pressure level& varies the metabolism, etc. Stress is the main cause of many diseases and more dangerous in long turn, but it is common and can be reduced with proper counselling and workouts.

Various factors of mental-health problems among the Indian farmers include changing life-styles, frequent income disruptions, crop failures, natural-calamity like flood & drought, economic-crisis, lacking in social-support, unemployment, and insecurity (Chatterjee, 2009). In India, since the mid-1990s, the suicide of farmers in various regions provides an illustration on mental-health due to financial-insecurity (McCurdy and Carroll, 2000; Mohanty, 2005; Sundar, 1999). In previous studies the socio-economic factors have been revealed to be contributing in stresses and worries among farmers that lead to their ultimate suicide-attempts (Behere and Behere, 2008; Dongre and Deshmukh, 2012). The farm activities are full of heavy physical work. The Indian farmers perform these activities in hot summer, heavy rain and cold winter. Farming processes are categorized in to different activities like Ploughing, harvesting, seeding, weeding etc.

Some of the activities are done manually and some with the help of machineries, all are performed in different procedure and with different equipment's. But almost all types of job are repetitive in nature and needs a lot of physical effort. Due to continuous physical workload the farmers faces a lot of pain in their knee, backbone, shoulder etc. With continuous physical stress farmer feels discomforts and pains in their body. Discomfort able body posture and work related injuries generates in the body also affect the mental condition of farmer. Socio economic problem and physical discomfort (health issues) creates biological changes in farmer's body and mind. Farmer unable to control or bear these significant changes and behavioral and emotional changes occur (i.e. anger, shyness, unhappy feeling, loneliness, anxiety, nervous, insecure, timid, forgetfulness etc). These feelings are nothing but the symbol of mental sickness found among farmers working in Agri field. The result of mental sickness is worst, it may decrease the working capacity, hamper health, generate chronic disease, reduce physical strength, increase suicidal syndrome.

Exposure to poor climatic conditions while performing agricultural tasks may cause occupational hazards that influence human health, reducing productivity, work quality, and increasing musculoskeletal-related issues. The stress also effects social, economic and family life of farmers.These stresses have a wide range of effects on an individual's well-being and health, and there has been a greater need to assess farmer stress levels so that adequate and appropriate preventive measures can be taken to improve farm productivity while also balancing the health of the farming community.

The farm activities are full of heavy physical work. The Indian farmers perform these activities in hot summer, heavy rain and cold winter. Farming processes are categorized in to different activities like Ploughing, harvesting, seeding, weeding etc. Some of the activities are done manually and some with the help of machineries, all are performed in different procedure and with different equipment's. But almost all types of job are repetitive in nature and needs a lot of physical effort. Due to continuous physical workload the farmers faces a lot of pain in their knee, backbone, shoulder etc. With continuous physical stress farmer feels discomforts and pains in their body. Discomfort able body posture and work related injuries generates in the body also affect the mental condition of farmer. Socio economic problem and physical discomfort (health issues) creates biological changes in farmer's body and mind. Farmer unable to control or bear these significant changes and behavioral and emotional changes occur (i.e. anger, shyness, unhappy feeling, loneliness, anxiety, nervous, insecure, timid, forgetfulness etc). These feelings are nothing but the symbol of mental sickness found among farmers working in Agri field. The result of mental sickness is worst, it may decrease the working capacity, hamper health, generate chronic disease, reduce physical strength, increase suicidal syndrome.

Exposure to poor climatic conditions while performing agricultural tasks may cause occupational hazards that influence human health, reducing productivity, work quality, and increasing musculoskeletal-related issues. The stress also effects social, economic and family life of farmers.These stresses have a wide range of effects on an individual's well-being and health, and there has been a greater need to assess farmer stress levels so that adequate and appropriate preventive measures can be taken to improve farm productivity while also balancing the health of the farming community

The farmers are supposed to be experiencing various stresses due to illiteracy, larger families, inappropriate labour-management, government regulations and corruptions, repayment of loans to banks & money-lender (Pande and Savenije, 2016), as well as indebtedness resulting their suicides (Merriott, 2016). Most of the agricultural activities are often characterized by manual-labour, long working hours, and higher level of stresses. The common occupational stressors among the farmers were reported as unfavourable environment-conditions, varying market-prices for crops as well as livestock, environmental-regulations, and time-pressures (Kearney et al., 2014). Ahearn (2011) has revealed the young farmers to be experiencing additional stressors that include limited-access to sites, capital, and affordable healthcare-facilities, and the difficulties to maintain an off-farm job and to establish them in such industry. Additionally, they may be more stressed as a function of managerial control as well as supports in comparison to the old age groups (Marotz-Baden and Mattheis, 1994; Weigel et al., 1987). Previous evidence suggested of chronic-stresses contributing in the development of mental-health illnesses, such as depressions and anxieties (Habib et al., 2001).

Different researchers have specified agriculture and its related activities to be a stressful-occupation (Fraser et al., 2005; Price and Evans, 2009; Roy et al., 2013). Because of the harder work in different challenging conditions, agriculture is usually associated with a variety of physical as well as mental-health risks (Fraser et al., 2005). The mental disorders like anxiety and depression are a reaction to the life stresses that causes a person in experiencing a long-term depressed state of mood with a depressive disorder and losing of interests in activities (Department of Health and Ageing, 2013), and further suicide-attempts among farmers (Liu et al., 2018). Moreover, the mental-illnesses among farmers may adversely affect their normal life a variety of ways like lesser interest in pleasures, smaller amount concentration, appetite loss, changing in weight, tiredness, tetchiness, sleeping problems, fatigues, reduced control levels, and anxieties (Walker and Walker, 1987; Simsek et al., 2015; Terrazas et al., 2018).

Different studies have also revealed that the agricultural workers have a higher tendency and prevalence of mental-illness as compared to the other non-farm workers in addition to other occupations like building-construction, teaching, and office-work (Walker and Walker, 1988; Kolstrup et al., 2008; Arcury et al., 2018; Rayens and Reed, 2014; Torske et al., 2015; Yin et al., 2018). The neural-systems are said to be affected by neurotoxic pesticides that cause mental-illnesses (Corral et al., 2017; Malekirad et al., 2013). The association with depression was found to be stronger among the farmers having earlier occurrences of pesticide poisoning than those with no such experience (Koh et al., 2017). The influencing factors for adaptive-decisions of farmers have been revealed as the availability in addition to admittance to services, resources, and technology (Antwi-Agyei et al., 2015). A significant risk factor for farmers' suicide was revealed as psychiatric problem, based on a study on farmers-households (Torske et al., 2016). The mental-illness of farmers owing to adverse impacts' of agriculture related activities that may lead to suicidal-attempts. (Yazd et al., 2019) have suggested of requirement of knowledge regarding the risk factors for farmers' psychological disorder and its impact in order to reduce the mental-illness burden. Further, it was recommended of the need of future research on impacts of climate change, farmers' mental health in developing country, and the information on reducing help seeking-barriers among farmers. Moreover, the most frequent injuries among the farmers that affected their mental-health were reported as musculoskeletal-disorders, cuts, and fractures (Rostamabadi et al., 2019).

The Canadian farmers reported of depression symptoms were 35% (Jones-Bitton et al., 2019). Bomble and Lhungdim (2020) have examined the mental-health status of farmers in Maharashtra (India) and found about 58% of farmers to be reported of distress of mental-health. the applications of 'multi-criteria decision making (MCDM)' methods have been increasing for various significant areas (Chaharsooghi and Ashrafi, 2014). As the decision-makers' judgments are sometimes uncertain and also, remain undefined with certainty. Therefore, it becomes difficulty in making the decision process. For the selection of best-suppliers in MCDM approach, an integrated 'grey theory and TOPSIS' model was suggested (Jadidi et al., 2008). Similarly, Aliakbarzadeh and Tabriz (2014) have applied "Fuzzy AHP and TOPSIS" for evaluating the performances in addition to the ranking of different branches banks. (Elsayed et al., 2017) have used the TOPSIS method with entropy-weight for ranking the banks based on their abilities to be competitive within the sectors.

Huang and Jiang (2018) have presented a model by defining an optimism-coefficient for the expansion of the physical-meaning of the standard TOPSIS that enable the decision-makers to describe their risk as well as profit related attitudes' through changes in the optimism-coefficient value. Similarly, the TOPSIS method has been widely used by different authors as well as researchers for a number of applications in past studies (Dang et al., 2019; He and Jung, 2018; Mitra and Kundu, 2018; Sari et al., 2018). Moreover, Quality Function Deployment (QFD) has been regarded as a planned-approach to define the customer-needs and for translating them into specific-plans in order to achieve the needs. This method has been an effective-communicating as well as quality-planning tool. Satapathy (2014) has used the "Artificial Neural-Network (ANN), QFD and "Interpretive Structural Modeling (ISM)" approach for the electric-utilities services in India for the consumers-satisfaction, and examined the inter-relationship between various design-requirements.

Lombardi et al., (2018) have used QFD method for assessing and prioritizing the risks in addition to hazards in a small-company producing wheat by using a wheeled cabin-tractor and other agricultural-equipment such as trailers, ploughs, harrows, and so on. Chakraborty et al., (2018) have used the QFD technique to calculate the priority-weights of the considered cotton-fibre properties in the textile-industries with the goal to resolve the issues with cotton-fibre selection.

Moreover, it has been reported in earlier studies that ISM enables in the management of interrelations among two or more elements/issues (Raj et al., 2008). For an effective description of a hierarchy among the variables, elements or influencers, ISM provides a useful framework for any project under concern (Sage, 1977; Warfield, 1974). A wide range of fields have found the suitability in ISM application that includes: analyzing the interactions between barriers in energy saving (Wang et al., 2008), analyzing the drivers and barriers that influence in implementing green supply-chain management (Mathiyazhagan et al., 2013), and analyzing the interactions between barriers in smart grid-technology (Luthra et al., 2014).

Moreover, a number of inter-related factors are said to be associated with any complex issue or problem under consideration, thus for a proper depiction of the situation, the direct as well as indirect relationships among those factors will be more appropriate as compared to individual factors (Mishra and Satapathy, 2020; Satapathy, 2014). Hence, in this study focus was made to find the most important parameters for mental stress among farmers of India. "Matrix Impact Cross-Reference Multiplication Analysis (MICMAC)" was done for the identification and clustering of the key risk factors. In the year 1973, MICMAC was developed by Duperrin and Godet (Wang et al., 2008), which was used in defining and analyzing the dependency as well as driving-force of variables (Mandal and Deshmukh, 1994). In MICMAC analysis, the factors have been classified in four-categories, such as "Autonomous, Linkage, Dependent and Driving factors" (Mishra and Satapathy, 2020; Ravi and Shankar, 2005). Indian farmers and the findings described the burden of depressions that may lead to the suicidal tendency among the farmers in Odisha and Chhattisgarh states in India. The present study revealed the contribution of farming to poor mental-health conditions (Ellis and Albrecht, 2017; Morgan et al., 2016). Based on the findings of this study, there should be immediate interventions initiations by the government in addition to the competent authorities with a view to mitigate the suicidal-attempts among the farmers because of depressions and mental-illnesses.

Physical stress occurs by any kind of injury, infection, surgery, over working, polluted environment or due to, sickness and musculoskeletal disorder. Psychological stress is called as emotional stress which occurs by fear, emotional factors or cognitive factor. Mental stress is the reaction to mind due to the effect of physical and psychological stress.The wellbeing of farmer is a major concern as they are expose to more physical work and repetitive work with different body postures.

Some of the farmers are stressed due to family and financial issues. So this becomes psychosocial stress that affects farmer community in a large scale.A few farmers commit suicide because they are unable to confront the issues that are expected to cause psychological stress.

The impact of mental-health issues on farmer health and production in agriculture has decreased over the world, which can have major effects not only for individuals, but also for entire agricultural sectors. Farmers' performance, illness rates, absenteeism in the workplace, farm-related injuries and accidents are all heavily influenced by mental stress. The farming community typically experiences stress when faced with demands that require them to change in some way; this increases psychological stress. Stress can affect an individual's ability to concentrate, as well as feelings of tension, apprehension, and panic, as well as physiological reactions such as perspiration, an increased heart rate, tensed muscles, and shallow breathing. Individuals who perceive the stressors to be more dangerous are more likely to experience higher-stress responses than those who can respond constructively to those stressors. The problems and stress have become so widespread that the farmers are unable to deal with them. Changes in climate, rainfall, crop disasters, strategies, government-plans, and, more importantly, if farmers are emotionally worried, it may result in a decrease in output rates and productivity.

India's economy is dependent on agriculture. More than 70% of the people are either farmer or dependent on farming. Maximum small farmers of Indian preferred to with conventional tools, in adverse climate. The nominal/small farmers having land below 1 or 2 hectare have to work alone and most of the lands are small/parted lands. Solitary work place, hot and humid climate, heavy work like ploughing, harvesting etc. make them mentally and physically stresses. A farmer has to work hard in their fields to grow good crops to feed their family members as well as society. Farmers face various uncontrollable pressures like natural calamities, uneven weathers, insect attacks, and economic pressures by Govt etc. which directly affects crop production. With addition to it more physical and repetitive work, use of traditional tools make their life more hazardous. Agricultural stress research is divided into three aspects: internal, external and professional. Acute stress related to internal agricultural needs (such as disease and adverse weather conditions) external sources of stress (such as skills and parameters) and uncertainty about capital and households are related to longer-lasting anxiety. Distinguish; the professional aspects of agricultural stress, such as heavy workload issues and agricultural practices, have a combination of significant and psychological impact increased paper work increased workload and changes in agricultural regulations. These have sparked an upsurge in activities deemed ineffective by the agricultural community and also they represent an important area of political participation. Major farmers showed the most obvious manifestations of pressure sometimes related to internal external and labour-management relations.

Due to this gap, families and temporary workers continue to lose their independence in working methods. The professional aspects of work and social pressure are an indispensable part of traditional agriculture.

It becomes a pre-requisite to consider the anthropometric dimensions of human body-parts in the initiation of any design activity. Jadav and Shah (2004) have found a significant correlation among head-length and body-height in Gujarat region. (Agarwal et. al., 2014) have found a gender difference with-respect to the mean cranial-lengths, cranial-breadth, which were initiate to be considerably greater in male as compare to female. The relationship coefficient among the measured cranial dimensions and stature were originated to be statistically significant as well as positive in together male and female. Further, it was concluded that by applying the regression equations other parameters can be known, if one of the parameter was known. Stature can be expressed as stable proportional relationship with other human-body parts measurements like face, head, trunk, and extremities etc. Marko et al., (2018) have conducted a study to estimate a correlation among height against head-breadth and vertex-height.

Moreover, the genetic-algorithm (GA) has been found to be useful for different optimisation problems applications by different authors and researchers. For instance, (Mardle et al., 2000) have used a non-linear goal-programme of the North-Sea demersal-fishery for developing a genetic-algorithm for optimisation. Further, in direct to calculate the relative effectiveness a comparison among the GA approach and conventional solution method was made. Genetic algorithms (GAs) is a heuristic-search and optimisation techniques encouraged by natural-evolution, which have been profitably applied to a large variety of real-world problem with significant complexities. McCall (2005) has used GA for a medical optimal-control problem. (Haldurai et al., 2016) have shown the combination of GA with different other methods as well as techniques to derive optimal-solutions that increased the computational time of retrieval-system.

Zhao and Kong (2016) have proposed an improved cyclic search genetic algorithm in order to overcome the shortcomings of early-convergences of genetic-algorithm and to resolve local minimization problems in search process. (Vaishnav et al., 2017) have made a comparison of different advantages and disadvantages of various algorithms for solving "Traveling Salesman Problem" by the use of GA. (Sekulić et al., 2018) have used a "response surface methodology (RSM)", "genetic-algorithm (GA)" and a "grey-wolf optimizer (GWO)" algorithm for surface-roughness prediction in ball-end-milling of hardened-steel. The predicted values of these models were compared with experimental results, and it was found that the best solution was GWO model. (Amjad et al., 2018) have made a review of literature on the use of GA to solve "Flexible Job Shop Scheduling Problems (FJSSP)".

Saeidi (2018) has proposed the meta-heuristic approach based on Simulated Annealing (SA) and Genetic Algorithm (GA) to solve the "Rubik's cube of 3D puzzles with 6 varying colored faces. It was found from the simulation result that the GA approach was more efficient to get shorter movement sequences than SA, while the convergence speeds as well as calculation time of the SA method was considerably lesser than GA. (Peng et al., 2019) have proposed a multi-agent genetic algorithm based on tabu search (MAGATS) in view of non-deterministic polynomial properties to solve job-shop scheduling-problems under make-span constraints. Gu et al., (2019) have proposed an "improved genetic-algorithm with adaptive-variable neighbourhood-search (IGA-AVNS)", which was found to be an effective algorithm to solve flexible job-shop scheduling problems. Eckroth and Schoen, (2019) have described the genetic-algorithm for achieving and selecting news-stories about artificial-intelligence for AAAI's weekly AI-Alert.

2. SUN FLOWER OPTIMIZATION (SFO) TECHNIQUE

The pattern of a sunflower is consistently the equivalent: consistently, they stir and go with the sun similar to the needle of a clock. Around evening time, they venture to every part of the other way to stand by again for their takeoff the following morning. Yang (2012) has projected another calculation dependent on the blossom fertilization procedure of blooming plants thinking about the organic procedure of multiplication. Right now, creators consider the impossible to miss conduct of sunflowers in the quest for the top direction toward the sun. The fertilization consider here was take haphazardly along the negligible separation among the bloom I and the blossom i+ 1. In reality, each bloom fix frequently discharge a huge number of dust gametes. In any case, for effortlessness, we likewise accept that every sunflower just creates one dust gamete and replicates exclusively.

Another significant nature-based enhancement here is about the opposite square law radiation. The law say that the force of the radiation is conversely corresponding to the square of the separation, i.e., the power (measure) of radiation diminishes with respect to the square of the expansion in separation. On the off chance that the separation copies, the power decreases by a factor 4, significantly increases, lessens to a factor 9, etc. For our situation, the less the good ways from the plant to the sun, the more noteworthy the measure of radiation got, and it will in general balance out in these regions. Then again, the more separation a plant is from the sun, the lesser measure of warmth got by it, so a similar will be follow right now will find a way to get as close as conceivable to the worldwide ideal (McInnes, 1999).

Subsequently, the quantity of heat Q received by every plant was given by:

$$Q_i = \frac{P}{4\pi r^2 i}$$

Here, P is the power of the sources and ri the distance between the present best and the plant i. The path of the sunflowers to the sun was:

$$\vec{s_i} = \frac{\dot{x} - x_i}{\left\| \dot{x} - x_i \right\|}, \ i = 1, 2, \ldots.., n_{p.}$$

The stage of the sunflowers on the direction s was designed by:

$$d_i = \lambda \times P_i (|| x_i - x_{i-1} ||) \times || x_i - x_{i-1} ||,$$

Where, λ is the steady worth that characterizes an "inertial" uprooting of the plants, Pi (|| xi - xi-1||) is the likelihood of fertilization, for example the sunflower I pollinates with its closest neighbor i– 1 producing another person in an arbitrary position that fluctuates as per each separation between the blossoms. That is, people nearer to the sun will make littler strides looking for a nearby refinement while increasingly inaccessible people will move ordinarily. It is additionally important to confine the greatest advance given by every person, all together not to skip locales inclined to be worldwide least competitors. Here, the most extreme advance was characterized as:

$$d_{max} = \frac{\left\| x_{max - x_{min}} \right\|}{2 \times N_{POP}}$$

Where, X max, and X min were the higher and inferior bound values, and N pop was the numeral of plants of total population. The new plantation will be:

$$\overrightarrow{X_{i+1}} = \vec{x_i} + d_i \times \vec{s_i}$$

The proposed calculation starts by the age of a populace of people. This populace might be arbitrary or even. The assessment of every individual permits assessing which one will be changed into the sun, that is, the one by the best assessment among all. Despite the fact that in a future variant, it is expected to embrace the chance of working with various suns, here right now is confined to just one. At that point, the various people will arrange themselves, similar to the sunflowers, toward the sun and will move arbitrarily forbidden, that is, they will make arbitrary strides a particular way.

3. METHODOLOGY

In this study four experts were considered from agricultural as well as design sectors to give their opinions. Then, the Fuzzy-TOPSIS method was used for the selection of most suitable hat material based on the opinion of experts' and the result of analysis. Further, based on the anthropometric head dimensions of farmers' head as well as heights, the "genetic algorithm (GA)" as well as "Sun Flower Optimization (SFO) method" were use to obtain the best fitness values for the fitness function for the height in association with other measured head parameters.

The steps followed in this work were as illustrated in figure .1.

Figure 1 shows the steps of the study in flow chart

The Fuzzy-TOPSIS method consisted of the following steps as illustrated below:

Step 1: In matrix format a classic fuzzy multiple attributes group decision-making problem was construct as:

$X_1\ X_2 \ldots X_n$

$$\hat{D} = \begin{matrix} A_1 \\ A_2 \\ \vdots \\ A_M \end{matrix} \begin{bmatrix} \tilde{x}_{11} & \tilde{x}_{12} & \cdots & \tilde{x}_{1n} \\ \tilde{x}_{21} & \tilde{x}_{22} & \vdots & \tilde{x}_{2n} \\ \vdots & \vdots & \ddots & \vdots \\ \tilde{x}_{m1} & \tilde{x}_{m2} & \vdots & \tilde{x}_{mn} \end{bmatrix}$$

$$\tilde{W} = (\tilde{W}_1, \tilde{W}_2, \ldots, \tilde{W}_n)$$

Figure 1. Steps in this study

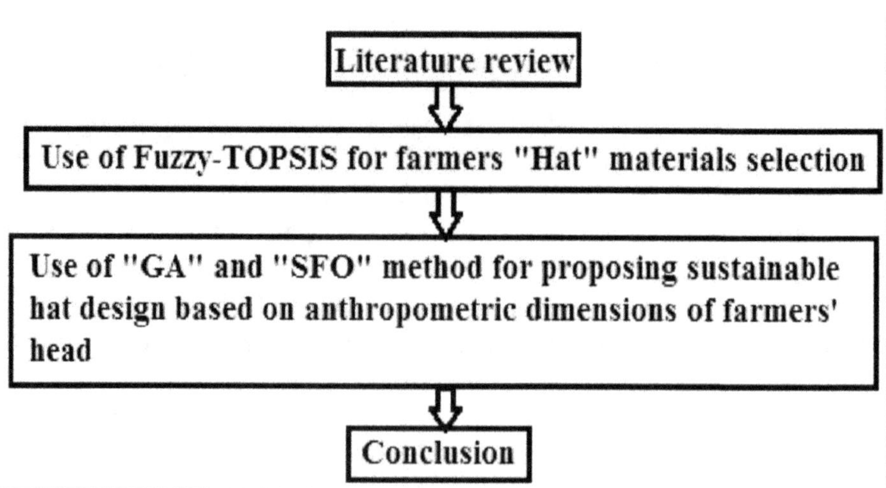

Here, A_1, A_2,..., A_m denoted the feasible alternative to be chosen, X_1, X_2..., X_n denoted the estimation attribute (criteria) measuring the alternative performances and correspond to the fuzzy performance rating of the ith alternative A_i opposed to the jth attribute X_j and $\tilde{W}_j$ denoted the attribute weight of X_j . Here, $\overline{x}_{ij}; \overline{\overline{x}}_{ij}, \forall_i, j$ = 1; 2;::: ; n were assess in linguistic term that were describe by triangular fuzzy number i.e.,

$$\overline{x}_{ij} = (a_{ij}, b_{ij}, c_{ij}), \tilde{W}_j = (w_{j1}, w_{j2}, w_{j3}).$$

Step 2: In order to evaluate the important weights of the attributes, a group of k experts was established such as:

$$\mathbf{E} = (E_1, E_2, ..., E_k)$$

In addition, each group member was assigned with different voting power weights according to their professional titles as:

$$\overline{\lambda} = (\overline{\lambda}_1, \overline{\lambda}_2, ...\overline{\lambda}_k)$$

Where, $\tilde{\lambda}_t$ represented the voting power weights of the t_{th} decisions maker that was expressed by triangular fuzzy numbers.

Step 3: The fuzzy group view matrix by all experts can be articulated as:

X_1 X_2 .... X_n

$$\tilde{w}^* = \begin{matrix} E_1 \\ E_2 \\ \vdots \\ E_M \end{matrix} \begin{bmatrix} \tilde{w}^*_{11} & \tilde{w}^*_{12} & \cdots & \tilde{w}^*_{1n} \\ \tilde{w}_{21} & \tilde{w}^*_{11} & \vdots & \tilde{w}^*_{11} \\ \vdots & \vdots & \ddots & \vdots \\ \tilde{w}^*_{k1} & \tilde{w}^*_{k2} & \vdots & \tilde{w}_{Kn} \end{bmatrix}$$

Where, $_{tj}$ indicated the fuzzy weight by the tth evaluator to the jth assessed attributes.

Step 4: The following equation was used to integrate all the experts' opinions in order to collective the individual judgement of k experts' to obtain the fuzzy weight ($\tilde{W}_j$) of criteria (X_j).

$$\tilde{W}_j = \sum_{t=1}^{k} \tilde{\lambda}_t \otimes \tilde{w}^{*}_{tj}, j = 1, 2, ..., n$$

Step 5: Then by apply the linear scale transformation process, the normalization of fuzzy decision matrix was performed as it preserve the property that the value of improved triangular fuzzy numbers remains in the range [0, 1]. Thus, the normalized fuzzy decision matrix ($\tilde{R}$) could be recognized as:

$$\tilde{R} = | \tilde{r}_{ij} | \mathbf{m \times n, i = 1, 2, . . ., m; j = 1,2, . . .,n}$$

$$\tilde{r}_{ij} = (a_{ij} / c_j^{+}, b_{ij} / c_j^{+}, c_{ij} / c_j^{+}), \mathbf{j}\ J^{+}$$

$$(a_j^{-} / c_{ij}, a_j^{-} / b_{ij}, a_j^{-} / c_{ij}), \mathbf{j}\ J^{-}$$

Where, $\{c_j^{+} = \max_i c_{ij}\ ,\ a_j^{-} = \max_i a_{ij},\ J^{+}\}$ was associated with benefit or maximization attributes or criteria and J^{-} was associated with cost or minimization attributes or criteria.

Step 6: Then with multiply the normalized fuzzy decision elements with the aggregative fuzzy weights of each criterion, the weighted normalized fuzzy decision matrix $\tilde{v}$ can be computed as:

$$\tilde{v} = | \tilde{v}_{ij} |_{m \times n},\ i = 1, 2, . . ., m, j = 1, 2, . . .,n.$$

Here, $\max_i \tilde{v}_{ij} = \tilde{w}_j \otimes \tilde{r}_{ij}$, and $\tilde{v}_{ij}$, $\forall_i$ j were positive triangular fuzzy numbers.

Step 7: Further, the fuzzy positive ideal solution (FPIS, A^{+}), as well as the fuzzy negative ideal solution (FNIS, A^{-}) were obtained as:

$$A^{+} = \{(\max_i \tilde{v}_{ij} | \mathbf{j}\ J^{+}), (\min_i \tilde{v}_{ij} | \mathbf{j}\ J^{-})\}$$

$$= \{\tilde{v}_1^{+}, \tilde{v}_2^{+}, . . ., \tilde{v}_j^{+}, ... \tilde{v}_n^{+},\}$$

$$A^{-} = \{(\min_i \tilde{v}_{ij} | \mathbf{j}\ J^{+}), (\max_i \tilde{v}_{ij} | \mathbf{j}\ J^{-})\}$$

$$= \{\tilde{v}_1^{-}, \tilde{v}_2^{-}, . . ., \tilde{v}_j^{-}, ... \tilde{v}_n^{-}\}$$

By considering the ranges of decision elements ($\tilde{v}_{ij}$, $\forall_i$ j) belonging to the closed interval [0, 1], it satisfied that

$$[\tilde{v}^{+}_{I+} = \tilde{\tilde{v}}^{-}_{I-} = (1,1,1)], \text{ and } [\tilde{\tilde{v}}^{-}_{j+} = \tilde{v}^{+}_{j-} = \tilde{v}^{+}_{j-} = (0;0;0)].$$

Where, J^{+} was associated with benefit criteria, and J^{-} was associated with cost criteria.

Step 8: Then, the distances of each alternative from (A^{+}) and (A^{-}) were obtained by the "Euclidean Distance Method" as:

$$S_{i+} = \sqrt{\sum_{j=1}^{n} d^2\left(\tilde{v}_{ij}, \tilde{\tilde{v}}^{+}_{j}\right)}, \ \mathbf{i = 1, 2, \ldots, m}$$

$$S_{i-} = \sqrt{\sum_{j=1}^{n} d^2\left(\tilde{v}_{ij}, \tilde{\tilde{v}}^{-}_{j}\right)}, \ \mathbf{i = 1, 2, \ldots, m}$$

Where, d ($\tilde{v}_A$, $\tilde{v}_B$) denoted the measured distance between two triangular fuzzy numbers i.e. $\tilde{A}$ and $\tilde{B}$.

Step 9: After calculating the (S_{i+}), and (S_{i-}) of each alternative, a closeness coefficient (c_i^{*}) was calculated to determine the final ranking order of all alternatives as:

$$c_i^{*} = s_{i-} / (s_{i+} + s_{i-}), \ 0 < c_i^{*} < 1$$

It may be noted that as (c_i^{*}) approaches to 1, when the alternative (A_i) gets closer to (A^{+}) and gets farther from (A^{-}), respectively. Therefore, according to the corresponding closeness coefficients, the ranking order of all alternatives can be obtained.

4 RESULTS AND DISCUSSION

Based on the opinion of experts', the four criteria chosen for the selection of hat materials were stiffness, cost, durability, and availability. Similarly, the alternative materials considered were Palm leaves, Plastic, Brown Straw, and Bamboo, respectively (Table .1).

Table 1 shows Selection of criteria and alternatives

Table 1. Selection of criteria and alternatives

Criteria	Symbol	Alternatives	Symbol
Stiffness	X_1	Palm leaves	A_1
Cost	X_2	Plastic	A_2
Durability	X_3	Brown Straw	A_3
Availability	X_4	Bamboo	A_4

The linguistic scale used for rating of the alternatives as defined by Chen and Hwang (1992) was as shown in Table 2.

Table 2 shows Linguistic scale with corresponding triangular fuzzy numbers

The criterions considered for maximization were stiffness, durability and availability. Whereas in this analysis, the criteria to be minimized was cost, The integrated matrix was obtained between the criterion and alternatives, with the corresponding weight of all the four criteria as shown in (Table .3), which was followed by the formulation of normalized matrix (Table .4), and weighted normalized matrix (Table .5). Further, the A+ and A- values were obtained between the criterion and alternatives (Table .6).

Table 3 shows Obtained integrated matrix

Table 4 shows Obtained normalized matrix

Table 5 shows Obtained weighted normalized matrix

Table 6 shows Calculated A^+ and A^- value

Similarly, the corresponding S^+ and S^- values for all the alternatives were as shown in Table X.7.

Table 7 shows Calculated S^+ and S^- values

Table 2. Linguistic scale with corresponding triangular fuzzy numbers

Linguistic scale	Symbol	Triangular Fuzzy Numbers
Very Poor	VP	0, 0, 1
Poor	P	0, 1, 3
Rather Poor	RP	1, 3, 5
Fair	F	3, 5, 7
Rather Good	RG	5, 7, 9
Good	G	7, 9, 10
Very Good	VG	9, 9, 10

Table 3. Obtained integrated matrix

	W_1			W_2			W_3			W_4		
	0.25	**0.46**	**0.78**	**0.77**	**0.82**	**1**	**0.67**	**0.86**	**0.92**	**0.77**	**0.93**	**1**
	X_1			X_2			X_3			X_4		
A_1	18.5000	25.2500	30.7500	3.0000	8.0000	14.7500	4.5000	10.0000	17.7500	3.5000	8.7500	15.7500
A_2	4.7500	10.2500	17.2500	13.5000	20.7500	27.7500	15.7500	23.2500	30.0000	10.2500	17.2500	24.5000
A_3	7.2500	13.5000	20.7500	4.2500	9.5000	16.2500	4.5000	10.2500	17.2500	3.0000	8.0000	14.7500
A_4	3.7500	8.5000	15.5000	23.7500	30.5000	35.0000	9.5000	16.2500	23.5000	24.2500	30.7500	35.0000
Max	18.5000	25.2500	30.7500	23.7500	30.5000	35.0000	15.7500	23.2500	30.0000	24.2500	30.7500	35.0000
Min	3.7500	8.5000	15.5000	3.0000	8.0000	14.7500	4.5000	10.0000	17.2500	3.0000	8.0000	14.7500

Table 4. Obtained normalized matrix

	W1			W2			W3			W4		
	0.25	**0.46**	**0.78**	**0.77**	**0.82**	**1**	**0.67**	**0.86**	**0.92**	**0.77**	**0.93**	**1**
	X_1			X_2			X_3			X_4		
A_1	0.6016	1.0000	1.6622	0.2034	1.0000	4.9167	0.1500	0.4301	1.1270	0.1000	0.2846	0.6495
A_2	0.1545	0.4059	0.9324	0.1081	0.3855	1.0926	0.5250	1.0000	1.9048	0.2929	0.5610	1.0103
A_3	0.2358	0.5347	1.1216	0.1846	0.8421	3.4706	0.1500	0.4409	1.0952	0.0857	0.2602	0.6082
A_4	0.1220	0.3366	0.8378	0.0857	0.2623	0.6211	0.3167	0.6989	1.4921	0.6929	1.0000	1.4433

Table 5. Obtained weighted normalized matrix

	X_1			X_2			X_3			X_4		
A_1	0.1504	0.4600	1.2965	0.1566	0.8200	4.9167	0.1005	0.3699	1.0368	0.0770	0.2646	0.6495
A_2	0.0386	0.1867	0.7273	0.0832	0.3161	1.0926	0.3518	0.8600	1.7524	0.2255	0.5217	1.0103
A_3	0.0589	0.2459	0.8749	0.1422	0.6905	3.4706	0.1005	0.3791	1.0076	0.0660	0.2420	0.6082
A_4	0.0305	0.1549	0.6535	0.0660	0.2151	0.6211	0.2122	0.6011	1.3727	0.5335	0.9300	1.4433

Table 6. Calculated A^+ and A^- values

	X_1			X_2			X_3			X_4		
A^+	0.1504	0.4600	1.2965	0.1566	0.8200	4.9167	0.3518	0.8600	1.7524	0.5335	0.9300	1.4433
A	0.0305	0.1549	0.6535	0.0660	0.2151	0.6211	0.1005	0.3791	1.0076	0.0660	0.2420	0.6082

The final ranking of alternatives (Table X.8) based on the closeness coefficient values indicated that the Bamboo material for the design of hat ranked first which was followed by the materials such as Plastic, Palm leaves, and Brown Straw, respectively.

Table 7. Calculated S^+ and S^- values

S_i^+	S1+	1.2686	S_i^-	S1-	0.9940
	S2+	1.1327		S2-	1.0528
	S3+	1.5907		S3-	0.6310
	S4+	1.1455		S4-	1.1202

Table 8 shows Ranking of Alternatives

Moreover, for a sustainable design of farmers' hat regardless of adverse climatic conditions, the anthropometric dimensions of head of selected 53 farmers were considered such as "Height, Head breadth, Face breadth, Face length, Menton to back of head, and Head length", respectively.

Thus, the obtained regression equation by the use of "Minitab 2017 version" was:

Height (cm) = [176 – (0.386 × Head breadth)] – [0.144 × (Face breadth)] – [0.109 × (Face length)] – [0.133 × (Menton to back of head)] + [0.113 × (Head length)]

Therefore, the fitness-function used for the corresponding analysis by "GA" and "SFO" was as follows:

z = (176 + (-0.386*X(1)) + (-0.144*X(2)) + (-0.109*X(3)) + (-0.133*X(4)) + 0.113*X(5));

As shown in Figure X.2 for the output of genetic algorithm (GA), it was observed of the best fitness value as 0.738533 with mean value as 0.73888.

Figure 2 shows Output of GA.

The MATLAB simulation worksheet of "Sun Flower Optimization" method used in this study was as illustrated in Figure .3. Further, based on the output obtained by the use of "Sun Flower Optimization" method as shown in Figure X.4, the "Total number of function evaluation (Sun Flower) = 2000", "Best solution (Sun Flower) = [16.5 15.2 13.5 20.7 17.5]", and "Best Fitness (Sun Flower) = 165.1951".

Table 8. Ranking of alternatives

Alternatives	Closeness coefficient value	Ranking
A_1	0.4393	3
A_2	0.4817	2
A_3	0.2840	4
A_4	0.4944	1

Figure 2. MATLAB simulation worksheet of "Sun Flower Optimization"

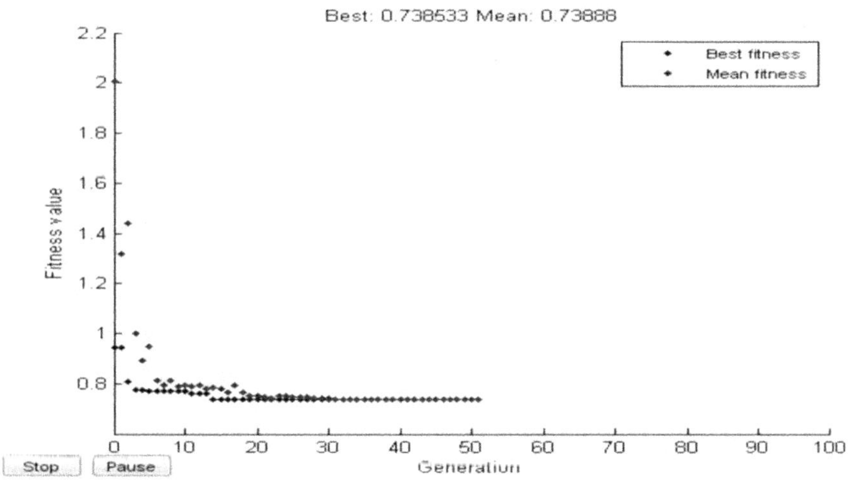

Figure 3. shows Output of "Sun Flower"

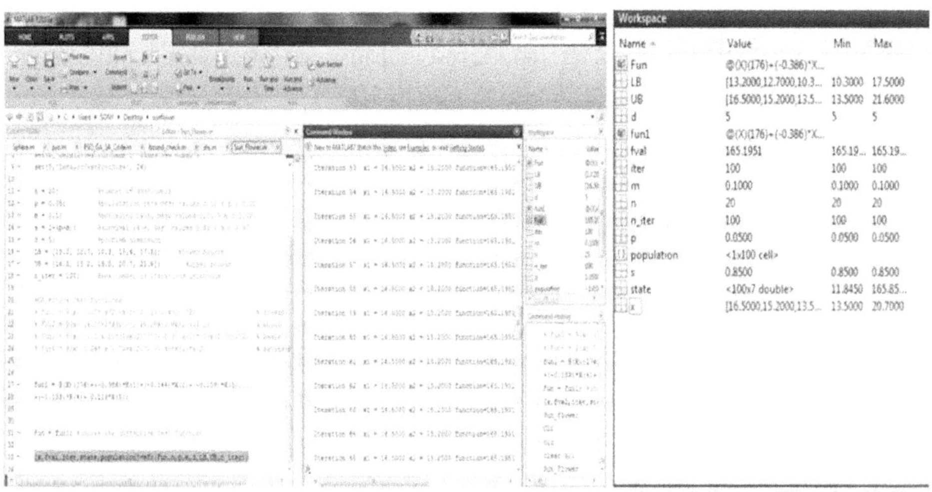

5. CONCLUSION

This study proposed a hybrid MCDM approach based on Fuzzy-TOPSIS method to evaluate the existing hat or cap materials for farmers' of India as a protective measure to their head in adverse climatic conditions. Four numbers of experts from the agricultural as well as design sectors were asked to rate the existing materials for farmers' cap based on some important criteria such as stiffness, cost, durability, and availability by using the linguistic scale to provide linguistic data. Then based on the Fuzzy-TOPSIS output, the Bamboo material for the design of hat ranked first which was followed by the materials such as Plastic, Palm leaves, and Brown Straw, respectively. Moreover, the proposed Fuzzy-TOPSIS method, genetic algorithm and sunflower optimization method provide an accurate, systematic, and efficient decision tool that enables the policy makers for making sound decisions. The result will provide an insight to further design the hat or cap in context with the Indian climatic conditions for nominal farmers'.

REFERENCES

Agarwal, S., Agarwal, S. K., & Jain, S. K. (2014). Correlation between the stature and cranial measurements in population of North India. *ActaMedica International*, *1*(2), 99–102. doi:10.5530/ami.2014.2.10

Ahearn, M. (2011). Potential challenges for beginning farmers and ranchers. In *CHOICES Online Magazine*. Agricultural & Applied Economics Association.

Aleixandre, J. L., Aleixandre-Tudo´, J. L., Bolan˜os-Pizarro, M., & Aleixandre-Benavent, R. (2015). Mapping the scientiðc research in organic farming: A bibliometric review. *Scientometrics*, *105*(1), 295–309. doi:10.100711192-015-1677-4

Aliakbarzadeh, A., & Tabriz, A. A. (2014). Performance evaluation and ranking the branches of bank using FAHP and TOPSIS case study: Tose asr shomal interest-free loan fund. *International Journal of Academic Research in Business & Social Sciences*, *4*(12), 199. doi:10.6007/IJARBSS/v4-i12/1339

Amjad, M. K., Butt, S. I., Kousar, R., Ahmad, R., Agha, M. H., Faping, Z., Anjum, N., & Asgher, U. (2018). Recent research trends in genetic algorithm based flexible job shop scheduling problems. *Mathematical Problems in Engineering*, *2018*, 2018. doi:10.1155/2018/9270802

Antwi-Agyei, P., Dougill, A. J., & Stringer, L. C. (2015). Barriers to climate change adaptation: Evidence from northeast Ghana in the context of a systematic literature review. *Climate and Development*, *7*(4), 297–309. doi:10.1080/17565529.2014.951013

Arcury, T. A., Sandberg, J. C., Talton, J. W., Laurienti, P. J., Daniel, S. S., & Quandt, S. A. (2018). Mental Health Among Latina Farmworkers and Other Employed Latinas in North Carolina. *Rural Mental Health*, *42*(2), 89–101. doi:10.1037/rmh0000091 PMID:30237844

Barman, S., Rouf, R. R., Islam, S., & Hasan, Md. M. (2017). *Proposed Design for Automated Agro-Quadcopter "Agro-ropter". 1st International Conference on Mechanical Engineering and Applied Science Military Institute of Science and Technology*, Dhaka, Bangladesh.

Behere, P. B., & Behere, A. P. (2008). Farmers' suicide in Vidarbha region of Maharashtra state: A myth or reality? *Indian Journal of Psychiatry*, *50*(2), 124. doi:10.4103/0019-5545.42401 PMID:19742218

Belton, V., & Stewart, T. (2002). *Multiple criteria decision analysis: An integrated approach*. Springer. doi:10.1007/978-1-4615-1495-4

Bian, X., Yao, G., & Shi, G. (2020). Social and natural risk factor correlation in China's fresh agricultural product supply. *PLoS One*, *15*(6), e0232836. doi:10.1371/journal.pone.0232836 PMID:32484814

Bomble, P., & Lhungdim, H. (2020). Mental health status of Farmers in Maharashtra, India: A study from farmer suicide prone area of Vidarbha region. *Clinical Epidemiology and Global Health*, *8*(3), 684–688. doi:10.1016/j.cegh.2020.01.002

Brigance, C., Soto Mas, F., Sanchez, V., & Handal, A. J. (2018). The mental health of the organic farmer: Psychosocial and contextual actors. *Workplace Health & Safety*, *66*(12), 606–616. doi:10.1177/2165079918783211 PMID:29962329

Çakmur, H. (2014). Health Risks Faced by Turkish Agricultural Workers. *TheScientificWorldJournal*, *2014*, 1–8. doi:10.1155/2014/185342 PMID:25054165

Cavanagh, J. T., Carson, A. J., Sharpe, M., & Lawrie, S. M. (2003). Psychological autopsy studies of suicide: A systematic review. *Psychological Medicine*, *33*(3), 395–405. doi:10.1017/S0033291702006943 PMID:12701661

Chaharsooghi, S. K., & Ashrafi, M. (2014). Sustainable supplier performance evaluation and selection with neofuzzy TOPSIS method. *International Scholarly Research Notices*, *2014*, 2014. doi:10.1155/2014/434168 PMID:27379267

Chakraborty, S., & Prasad, K. (2018). A quality function deployment-based expert system for cotton fibre selection. *Journal of The Institution of Engineers (India): Series E*, *99*(1), 43-53.

Chatterjee, P. (2009). Economic crisis highlights mental health issues in India. *Lancet*, *373*(9670), 1160–1161. doi:10.1016/S0140-6736(09)60666-8 PMID:19350704

Chen, S. J., & Hwang, C. L. (1992). *Fuzzy Multiple Attribute Decision Making Methods. Lecture notes in economics and mathematical system series* (Vol. 375). Springer.

Chico, B. (2013). *Hats and Headwear around the World: A Cultural Encyclopedia*. ABC-CLIO.

Corral, S. A., de Angel, V., Salas, N., Zúñiga-Venegas, L., Gaspar, P. A., & Pancetti, F. (2017). Cognitive impairment in agricultural workers and nearby residents exposed to pesticides in the Coquimbo Region of Chile. *Neurotoxicology and Teratology*, *62*, 13–19. doi:10.1016/j.ntt.2017.05.003 PMID:28579518

Dang, V. T., Wang, J., & Dang, W. V. T. (2019). An integrated fuzzy AHP and fuzzy TOPSIS approach to assess sustainable urban development in an emerging economy. *International Journal of Environmental Research and Public Health*, *16*(16), 2902. doi:10.3390/ijerph16162902 PMID:31412685

Das, A. (2011). Farmers' suicide in India: Implications for public mental health. *The International Journal of Social Psychiatry*, *57*(1), 21–29. doi:10.1177/0020764009103645 PMID:21252353

Deccan-herald. (2020). India may become farmer suicide capitol. *Decan Harold.* https://www.deccanherald.com/national/india-may-become-farmer-suicide-capital-agri-adviser-773068.html

Department of Health and Ageing. (2013). National Mental Health Report 2013: Tracking Progress of Mental Health Reform in Australia 1993–2011. Commonwealth of Australia: Canberra, Australia.

Dierk, C., Carter, S., Chiu, P., Dunnigan, T., & Kimber, D. (2019). Use Your Head! Exploring Interaction Modalities for Hat Technologies. *Wearables*, 1033-1045. 10.1145/3322276.3322356

Dongre, A. R., & Deshmukh, P. R. (2012). Farmers' suicides in the Vidarbha region of Maharashtra, India: A qualitative exploration of their causes. *Journal of Injury & Violence Research*, *4*(1), 2–7. doi:10.5249/jivr.v4i1.68 PMID:21502794

Dutta, M. (2018). If this is our land, where are our stories? Cultural piracy of crafts of Assam. *Locating Northeast India: Human Mobility, Resource Flows, and Spatial Linkages. Setting the agenda for Traditional Knowledge utilisation International Conference*, Tezpur University, India.

Eckroth, J.,& Schoen, E. (2019). *A Genetic Algorithm for Finding a Small and Diverse Set of Recent News Stories on a Given Subject: How We Generate AAAI's AI-Alert.*

Ellis, N. R., & Albrecht, G. A. (2017). Climate change threats to family farmers' sense of place and mental wellbeing: A case study from the Western Australian wheat belt. *Social Science & Medicine*, *175*, 161–168. doi:10.1016/j.socscimed.2017.01.009 PMID:28092757

Elsayed, E. A., Dawood, A. S., & Karthikeyan, R. (2017). Evaluating alternatives through the application of TOPSIS method with entropy weight. *Int. J. Eng. Trends Technol*, *46*(2), 60–66. doi:10.14445/22315381/IJETT-V46P211

Fraser, C. E., Smith, K. B., Judd, F., Humphreys, J. S., Fragar, L. J., & Henderson, A. (2005). Farming and mental health problems and mental illness. *The International Journal of Social Psychiatry*, *51*(4), 340–349. doi:10.1177/0020764005060844 PMID:16400909

Gregoire, A. (2002). The mental health of farmers. *Occupational Medicine (Philadelphia, Pa.)*, *52*(8), 471–476. PMID:12488518

Grzywacz, J. G., Quandt, S. A., Chen, H., Isom, S., Kiang, L., Vallejos, Q., & Arcury, T. A. (2010). Depressive Symptoms Among Latino Farmworkers Across the Agricultural Season: Structural and Situational Influences. *Cultural Diversity & Ethnic Minority Psychology*, *16*(3), 335–343. doi:10.1037/a0019722 PMID:20658876

Gu, X., Huang, M., & Liang, X. (2019). An Improved Genetic Algorithm with Adaptive Variable Neighborhood Search for FJSP. *Algorithms*, *12*(11), 243. doi:10.3390/a12110243

Habib, K. E., Gold, P. W., & Chrousos, G. P. (2001). Neuroendocrinology of stress. *Endocrinology and Metabolism Clinics of North America*, *30*(3), 695–728. doi:10.1016/S0889-8529(05)70208-5 PMID:11571937

Haldurai, L., Madhubala, T., & Rajalakshmi, R. (2016). A Study on Genetic Algorithm and its Applications. *International Journal on Computer Science and Engineering*, *4*, 139–143.

He, Y., & Jung, H. (2018). A voting TOPSIS approach for determining the priorities of areas damaged in disasters. *Sustainability (Basel)*, *10*(5), 1607. doi:10.3390u10051607

Huang, Y., & Jiang, W. (2018). Extension of TOPSIS Method and its Application in Investment. *Arabian Journal for Science and Engineering*, *43*(2), 693–705. doi:10.100713369-017-2736-3

Jadav, H. R., & Shah, G. V. (2004). Determination of personal height from the length of head in Gujarat region. *Journal of the Anatomical Society of India*, *53*(1), 20–21.

Jadidi, O., Hong, T. S., Firouzi, F., Yusuff, R. M., & Zulkifli, N. (2008). TOPSIS and fuzzy multi-objective model integration for supplier selection problem. *Journal of Achievements in Materials and Manufacturing Engineering*, *31*(2), 762–769.

Jadidi, O., Hong, T. S., Firouzi, F., Yusuff, R. M., & Zulkifli, N. (2008). TOPSIS and fuzzy multi-objective model integration for supplier selection problem. *Journal of Achievements in Materials and Manufacturing Engineering*, *31*(2), 762–769.

Jallow, M. F. A., Awadh, D. G., Albaho, M. S., Devi, V. Y., & Thomas, B. M. (2017). Pesticide Knowledge and Safety Practices among Farm Workers in Kuwait: Results of a Survey. *International Journal of Environmental Research and Public Health*, *14*(4), 340. doi:10.3390/ijerph14040340 PMID:28338612

Jazi, M. S., Rezaei, A., Karami, G., Azarmi, F., & Ziejewski, M. (2014). A computational study of influence of helmet padding materials on the human brain under ballistic impacts. *Computer Methods in Biomechanics and Biomedical Engineering*, *17*(12), 1368–1382. doi:10.1080/10255842.2012.748755 PMID:23281845

Johnson, R. A., & Wichern, D. W. (2002). Applied multivariate statistical analysis: Vol. 5. *No. 8*. Prentice hall.

Jones-Bitton, A., Best, C., MacTavish, J., Fleming, S., & Hoy, S. (2019). Stress, anxiety, depression, and resilience in Canadian farmers. *Social Psychiatry and Psychiatric Epidemiology*. doi:10.100700127-019-01738-2 PMID:31197397

Kallioniemi, M. K., Simola, A., Kaseva, J., & Kymäläinen, H.-R. (2016). Stress and Burnout Among Finnish Dairy Farmers. *Journal of Agromedicine*, *21*(3), 259–268. doi:10.1080/1059924X.2016.1178611 PMID:27081893

Kearney, G. D., Rafferty, A. P., Hendricks, L. R., Allen, D. L., & Tutor-Marcom, R. (2014). A cross sectional study of stressors among farmers in Eastern North Carolina. *North Carolina Medical Journal*, *75*(6), 384–392. doi:10.18043/ncm.75.6.384 PMID:25402689

Khan, I., Shinde, A., Khandelwal, S., Sah, V., & Narendiranath, B. T. (2019). FINITE ELEMENT Analysis On Different Materials For Motorcycle Helmet Shell. *International Research Journal of Engineering and Technology*, *6*(5), 4581–4586.

Koh, S.-B., Kim, T. H., Min, S., Lee, K., Kang, D. R., & Choi, J. R. (2017). Exposure to pesticide as a risk factor for depression: A population-based longitudinal study in Korea. *Neurotoxicology*, *62*, 181–185. doi:10.1016/j.neuro.2017.07.005 PMID:28720389

Kolstrup, C., Lundqvist, P., & Pinzke, S. (2008). Psychosocial Work Environment Among Employed Swedish Dairy and Pig Farmworkers. *Journal of Agromedicine*, *13*(1), 23–36. doi:10.1080/10599240801986157 PMID:19042690

Kou, G., Miettinen, K., & Shi, Y. (2011). Multiple criteria decision making: Challenges and advancements. *Journal of Multi-Criteria Decision Analysis*, *18*(1-2), 1–4. doi:10.1002/mcda.476

Latifi, S., Hauser, M., Raheli, H., Moghaddam, S. M., Viira, A.-H., Ozuyar, P. G., & Azadi, H. (2020). Impacts of organizational arrangements on conservation agriculture: Insights from interpretive structural modeling in Iran. *Agroecology and Sustainable Food Systems*. doi:10.1080/21683565.2020.1751375

Liu, B.-P., Qin, P., Liu, Y.-Y., Yuan, L., Gu, L.-X., & Jia, C.-X. (2018). Mental disorders and suicide attempt in rural China. *Psychiatry Research*, *261*, 190–196. doi:10.1016/j.psychres.2017.12.087 PMID:29309959

Lombardi, M., & Fargnoli, M. (2018). Prioritization of hazards by means of a QFD-based procedure. *Safety and Security Studies*, 163.

Luthra, S., Kumar, S., Kharb, R., Ansari, M. F., & Shimm, S. L. (2014). Adoption of smart grid technologies: An analysis of interactions among barriers. *Renewable & Sustainable Energy Reviews*, *33*, 554–565. doi:10.1016/j.rser.2014.02.030

Makalew, J., & Kusuma, Z., Sugiyanto, Tamod, Z. (2013). The Influence of Farmer Characteristic and Farming to the Farmer Motivation on Using Organic Manure (Case Study in East Tombatu District, South-East Minahasa Regency). *International Journal of Engineering Inventions*, *3*(1), 43–51.

Malekirad, A. A., Faghih, M., Mirabdollahi, M., Kiani, M., Fathi, A., & Abdollahi, M. (2013). Neurocognitive, mental health, and glucose disorders in farmers exposed to Organophosphorus pesticides. *(Report). Arhiv za Higijenu Rada i Toksikologiju*, *64*(1), 1–8. doi:10.2478/10004-1254-64-2013-2296 PMID:23705196

Mandal, A., & Deshmukh, S. G. (1994). Vendor selection using interpretive structural modeling (ISM). *International Journal of Operations & Production Management*, *14*(6), 52–59. doi:10.1108/01443579410062086

Mardle, S. J., Pascoe, S., & Tamiz, M. (2000). An investigation of genetic algorithms for the optimization of multi objective fisheries bioeconomic models. *International Transactions in Operational Research*, *7*(1), 33–49. doi:10.1111/j.1475-3995.2000.tb00183.x

Marko, R. S., Awasthi, A., Jehan, M., & Bose, A. (2018). A Correlation Analysis for Estimation of Height from Head Breadth and Vertex Height. *IOSR Journal of Dental and Medical Sciences*, *17*(4), 87–91.

Marotz-Baden, R., & Mattheis, C. (1994). Daughters-in-law and stress in two generation farm families. *Family Relations*, *43*(2), 132–137. doi:10.2307/585314

Mathiyazhagan, K., Govindan, K., NoorulHaq, A., & Geng, Y. (2013). An ISM approach for the barrier analysis in implementing green supply chain management. *Journal of Cleaner Production*, *47*, 283–297. doi:10.1016/j.jclepro.2012.10.042

McCall, J. (2005). Genetic algorithms for modelling and optimisation. *Journal of Computational and Applied Mathematics*, *184*(1), 205–222. doi:10.1016/j.cam.2004.07.034

McCurdy, S. A., & Carroll, D. J. (2000). Agricultural injury. *American Journal of Industrial Medicine*, *38*(4), 463–480. doi:10.1002/1097-0274(200010)38:4<463::AID-AJIM13>3.0.CO;2-N PMID:10982988

McInnes, C. R. (1999). Solar radiation pressure. In *Solar sailing* (pp. 32–55). Springer. doi:10.1007/978-1-4471-3992-8_2

McShane, C. J., Quirk, F., & Swinbourne, A. (2016). Development and validation of a work stressor scale for Australian farming families. *The Australian Journal of Rural Health*, *24*(4), 238–245. doi:10.1111/ajr.12261 PMID:26689163

Meeker, B. J., Carruth, A., & Holland, C. B. (2002). Health Hazards and Preventive Measures of Farm Women. *AAOHN Journal*, *50*(7), 307–314. doi:10.1177/216507990205000708 PMID:12141093

Merriott, D. (2016). Factors associated with the farmer suicide crisis in India. *Journal of Epidemiology and Global Health*, *6*(4), 217–227. doi:10.1016/j.jegh.2016.03.003 PMID:27080191

Mishra, D., & Satapathy, S. (2020). An Integrated SWARA, QFD and ISM approach for Agricultural injuries in India. *International Journal of Decision Support System Technology*, *12*(2), 1–24. doi:10.4018/IJDSST.2020040101

Mitra, S., & Kundu, S. (2018). Application of TOPSIS for best domestic refrigerator selection. *International Journal of Research and Analytical Reviews*, *5*(3), 226–231.

Mohanty, B. B. (2005). 'We are like the living dead': Farmer suicides in Maharashtra, western India. *The Journal of Peasant Studies*, *32*(2), 243–276. doi:10.1080/03066150500094485

Mora, D. C., Quandt, S. A., Chen, H., & Arcury, T. A. (2016). Associations of Poor Housing with Mental Health Among North Carolina Latino Migrant Farmworkers. *Journal of Agromedicine*, *21*(4), 327–334. doi:10.1080/1059924X.2016.1211053 PMID:27409300

Morgan, M. I., Hine, D. W., Bhullar, N., Dunstan, D. A., & Bartik, W. (2016). Fracked: Coal seam gas extraction and farmers mental health. *Journal of Environmental Psychology*, *47*, 22–32. doi:10.1016/j.jenvp.2016.04.012

Nath, B. (2019). Assamese Traditional Icon "Japi": History And Diversity. *RESEARCH REVIEW International Journal of Multidisciplinary*, *4*(8), 29–33.

Othman, A., & Owen, L. (2001). The multi dimensionality of Carter Model to measure customer service quality (SQ) in Islamic banking industry: A study in Kuwait finance house. *International Journal of Islamic Financial Services*, *3*(4), 1–12.

Padhy, C., & Raju, P. S. (2020). Mental Health of Farmers–Need of the Hour. *International Journal of Agriculture Environment and Biotechnology*, *13*(1), 7–91. doi:10.30954/0974-1712.1.2020.11

Pande, S., & Savenije, H. H. (2016). A sociohydrological model for smallholder farmers in Maharashtra, India. *Water Resources Research*, *52*(3), 1923–1947. doi:10.1002/2015WR017841

Peng, C., Wu, G., Liao, T. W., & Wang, H. (2019). Research on multi-agent genetic algorithm based on tabu search for the job shop scheduling problem. *PLoS One*, *14*(9), e0223182. doi:10.1371/journal.pone.0223182 PMID:31560722

Price, L., & Evans, N. (2009). From stress to distress: Conceptualizing the British family farming patriarchal way of life. *Journal of Rural Studies*, *25*(1), 1–11. doi:10.1016/j.jrurstud.2008.03.008

Raj, T., Shankar, R., & Suhaib, M. (2008). An ISM approach for modeling the enablers of flexible manufacturing system: The case for India. *International Journal of Production Research*, *46*(24), 6883–6912. doi:10.1080/00207540701429926

Ravi, V., & Shankar, R. (2005). Analysis of interactions among the barriers of reverse logistics. *Technological Forecasting and Social Change*, *72*(8), 1011–1029. doi:10.1016/j.techfore.2004.07.002

Rayens, M. K., & Reed, D. B. (2014). Predictors of Depressive Symptoms in Older Rural Couples: The Impact of Work, Stress and Health. *The Journal of Rural Health*, *30*(1), 59–68. doi:10.1111/jrh.12028 PMID:24383485

Rostamabadi, A., Jahangiri, M., Naderi Mansourabadi, B., Javid, M., Ghorbani, M., & Banaee, S. (2019). Prevalence of chronic diseases and occupational injuries and their inñuence on the health-related quality of life among farmers working in small-farm enterprises. *Journal of Agromedicine*, *24*(3), 248–256. doi:10.1080/1059924X.2019.1592047 PMID:30880615

Roy, P., Tremblay, G., Oliffe, J. L., Jbilou, J., & Robertson, S. (2013). Male farmers with mental health disorders: A scoping review. *The Australian Journal of Rural Health*, *21*(1), 3–7. doi:10.1111/ajr.12008 PMID:23384130

Saeidi, S. (2018). Solving the Rubik's Cube using Simulated Annealing and Genetic Algorithm. *International Journal of Education and Management Engineering*, *8*(1), 1–10. doi:10.5815/ijeme.2018.01.01

Safety and health in agriculture (2011). *ILO code of practice.* International Labour Office - Geneva: ILO.

Sage, A. (1977). *Interpretive structural modeling: Methodology for large-scale systems*. McGraw-Hill.

Saikia, S.B. (2012). *Design Audit Report.* Mukhkushi Japi making Cluster, 2012.

Sari, R. M., Rizkya, I., Syahputri, K., & Siregar, I. (2018). Alternative of raw material's suppliers using TOPSIS method in chicken slaughterhouse industry. *IOP Conference Series. Materials Science and Engineering*, *309*(1), 012041. doi:10.1088/1757-899X/309/1/012041

Satapathy, S. (2014). ANN, QFD and ISM approach for framing electricity utility service in India for consumer satisfaction. *International Journal of Services and Operations Management*, *18*(4), 404–428. doi:10.1504/IJSOM.2014.063243

Scarth, R. D., Stallones, L., Zwerling, C., & Burmeister, L. F. (2000). The prevalence of depressive symptoms and risk factors among Iowa and Colorado farmers. *American Journal of Industrial Medicine*, *37*(4), 382–389. doi:10.1002/(SICI)1097-0274(200004)37:4<382::AID-AJIM8>3.0.CO;2-4 PMID:10706750

Scheyett, A., Bayakly, R., & Whitaker, M. (2019). Characteristics and contextual stressors in farmer and agricultural worker suicides in Georgia from 2008–2015. *Rural Mental Health*, *43*(2-3), 61–72. doi:10.1037/rmh0000114

Sekulic, M., Pejic, V., Brezocnik, M., Gostimirović, M., & Hadzistevic, M. (2018). Prediction of surface roughness in the ball-end milling process using response surface methodology, genetic algorithms, and grey wolf optimizer algorithm. *Advances in Production Engineering & Management*, *13*(1), 18–30. doi:10.14743/apem2018.1.270

Shen, L., Song, X., Wu, Y., Liao, S., & Zhang, X. (2016). Interpretive structural modeling based factor analysis on the implementation of emission trading system in the Chinese building sector. *Journal of Cleaner Production*, *127*, 214–227. doi:10.1016/j.jclepro.2016.03.151

Simsek, Z., Ersin, F., & Kirmizitoprak, E. (2015). Development of the Seasonal Migrant Agricultural Worker Stress Scale in Sanliurfa, Southeast Turkey. *Journal of Agromedicine*, *21*(1), 56–60. doi:10.1080/1059924X.2015.1106372 PMID:26479204

Stallones, L., & Beseler, C. (2002). Pesticide illness, farm practices, and neurological symptoms among farm residents in Colorado. *Environmental Research*, *90*(2), 89–97. doi:10.1006/enrs.2002.4398 PMID:12483798

Staniford, A. K., Dollard, M. F., & Guerin, B. (2009). Stress and help-seeking for drought-stricken citrus growers in the Riverland of South Australia. *The Australian Journal of Rural Health*, *17*(3), 147–154. doi:10.1111/j.1440-1584.2009.01059.x PMID:19469779

Sundar, M. (1999). Suicide in farmers in India. *The British Journal of Psychiatry*, *175*(6), 585–586. doi:10.1192/bjp.175.6.585c PMID:10789361

Terrazas, S. R., & McCormick, A. (2018). Coping Strategies That Mitigate Against Symptoms of Depression Among Latino Farmworkers. *Hispanic Journal of Behavioral Sciences*, *40*(1), 57–72. doi:10.1177/0739986317752923

Torske, M. O., Hilt, B., Bjørngaard, J. H., Glasscock, D., & Krokstad, S. (2015). Disability pension and symptoms of anxiety and depression: A prospective comparison of farmers and other occupational groups. The HUNT Study, Norway. *BMJ Open*, *2015*(11), 5. doi:10.1136/bmjopen-2015-009114 PMID:26525724

Torske, M. O., Hilt, B., Glasscock, D., Lundqvist, P., & Krokstad, S. (2016). Anxiety and depression symptoms among farmers: The HUNT study, Norway. *Journal of Agromedicine*, *21*(1), 24–33. doi:10.1080/1059924X.2015.1106375 PMID:26488439

Vaishnav, P., Choudhary, N., & Jain, K. (2017). Traveling salesman problem using genetic algorithm: A survey. *International Journal of Scientific Research in Computer Science. Engineering and Information Technology*, *2*(3), 105–108.

Walker, J. L., & Walker, L. J. S. (1988). Self-reported stress symptoms in farmers. *Journal of Clinical Psychology*, *44*(1), 10–16. doi:10.1002/1097-4679(198801)44:1<10::AID-JCLP2270440103>3.0.CO;2-6 PMID:3343357

Walker, L. S., & Walker, J. L. (1987). Stressors and symptoms predictive of distress in farmers. *Family Relations*, *36*(4), 374–378. doi:10.2307/584486

Wang, G. H., Wang, Y., & Zhao, T. (2008). Analysis of interactions among the barriers to energy saving in China. *Energy Policy, 36*(6), 1879–1889. doi:10.1016/j.enpol.2008.02.006

Warfield, J. W. (1974). Developing interconnected matrices in structural modelling, IEEE transcript on systems. *Men and Cybernetics, 4*, 51–81.

Wästerlund, D. S. (2018). Managing heat in agricultural work: increasing worker safety and productivity by controlling heat exposure. *Forestry Working Paper No. 1*. Rome, FAO.

Weigel, R. R., Weigel, D. J., & Blundall, J. (1987). Stress, coping, and satisfaction: Generational differences in farm families. *Family Relations, 36*(1), 45–48. doi:10.2307/584646

Weisskopf, M. G., Moisan, F., Tzourio, C., Rathouz, P. J., & Elbaz, A. (2013). Pesticide Exposure and Depression Among Agricultural Workers in France. *American Journal of Epidemiology, 178*(7), 1051–1058. doi:10.1093/aje/kwt089 PMID:23851580

Yang, X.-S. (2012). *Flower pollination algorithm for global optimization*. In: *International conference on unconventional computing and natural computation*. Springer, Berlin. 10.1007/978-3-642-32894-7_27

Yazd, S. D., Wheeler, S. A., & Zuo, A. (2019). Key Risk Factors Affecting Farmers' Mental Health: A Systematic Review. *International Journal of Environmental Research and Public Health, 16*(23), 4849. doi:10.3390/ijerph16234849 PMID:31810320

Yin, H., Xu, G., Tian, H., Yang, G., Wardenaar, K. J., & Schoevers, R. A. (2018). The prevalence, age-of-onset and the correlates of DSM-IV psychiatric disorders in the Tianjin Mental Health Survey (TJMHS). *Psychological Medicine, 48*(3), 473–487. doi:10.1017/S0033291717001878 PMID:28714421

Zhao, H., & Kong, F. (2016). Research and Applications of Shop Scheduling Based on Genetic Algorithms. *Brazilian Archives of Biology and Technology, 59*(SPE.

Zhou, P., Fan, L. W., & Zhou, D. Q. (2010). Data aggregation in constructing composite indicators: A perspective of information loss. *Expert Systems with Applications, 37*(1), 360–365. doi:10.1016/j.eswa.2009.05.039

Zulqarnain, M., & Dayan, F. (2017). Choose Best Criteria for Decision Making Via Fuzzy Topsis Method. *Mathematics and Computer Science, 2*(6), 113–119. doi:10.11648/j.mcs.20170206.14

Chapter 9
Smart Crop Protection System From Wild Animals Using Artificial Intelligence

Shailaja S. Mudengudi
https://orcid.org/0000-0001-6102-7942
Tontadrya College of Engineering, India

Muktha S. Patil
Tontadarya College of Engineering, India

Neetha S. Mudaraddi
Tontadarya College of Engineering, India

ABSTRACT

The first major threat to the farmers is drought. Crop vandalization by animals is the second major threat after drought. Crops are vulnerable to animals. Therefore, it is very important to monitor the nearby presence of animals. The main aim of this project is to provide a better solution in order to resolve this problem. In this chapter, the authors proposed a method which could detect the presence of animal and offer a warning. They used a microcontroller and camera to detect the movement of animals, and send signal to the controller. It diverts the animal by producing sound; and, further, a signal is transmitted to GSM which gives an alert to the owner of the crop immediately. The proposed monitoring scheme is to provide an early warning about possible intrusion and damage by animals.

1. INTRODUCTION

Economy of many countries is majorly dependent on agriculture. Agriculture plays a major role in the economic development of a country. Agriculture is the main stay of

DOI: 10.4018/978-1-6684-9975-7.ch009

economy which contributes for the gross domestic product. Agriculture is the major source to meet the food essentials and requirements of the people. It also provides raw materials to the industries. But due to animal interference and fire incidents in agricultural lands, huge loss of crops is experienced by the farmers. To avoid the financial losses, it is very crucial to protect the agricultural fields and farms from animals. Our proposed work we have designed a system which prevents the entry of animals inside the farm by using LDR and Camera. The purpose of the designed framework is to develop intruder alert to the farm, so as to avoid the losses incurred due to animals and fire.

These alerts given by the proposed framework, protects the crop from damaging which in turn increase yield of the crop. The developed system will not harm or injure the animals as well as human beings in any form. Aim of the presented framework, is to design an intelligent security system which protects the farms with the aid of Embedded system which is based on Arduino as shown in Figure 1,2 &3. Due to the lack of of any detection system, these attacks kill villagers and destroy their crops as well. The villagers are left helpless to their fate due to absence of security measures. Therefore, a need for a proper detection system is raised, which could help and save their lives and also preserve the crops.

Several crops & paddy fields cannot always be fenced.

With the best utilization of mobile communication technology, the main objective of this paper is to utilize global system for mobile communication (GSM) and use short message service (SMS). This system helps us to provide surveillance functionality and keep away wild animals from the farm.

Several solutions followed by the framers so as to minimize the loss of crops is discussed next. It is found that the smell of rotten eggs prevents the wild pigs & deer from entering in to the farm and destroying the crops. The farmers use the rotten egg solution spray on their fields to prevent the entry of pigs and deer into the farm. Firecrackers are used to frighten and ward off the elephant.

The framework, along with providing protection from wild animals, distinguishes between an intruder and authorized person with the aid of RFID's and various LDR sensors.

These are deployed in the frame area to sense and detect any motion and turn's ON the camera when movement is detected, thereby providing real time monitoring.

Several methods are automated to suppress and prevent the wild animals from entering the farmlands and destroying the crops, such as electronic fire a rotten egg spray, to ward off the wild animals. The framework employs Haar feature which is based cascade classifiers to detect the object and distinguish between animal and human.

a) Animals coming near the field
b) Sound alert for animal entry

Figure 1.

Figure 2.

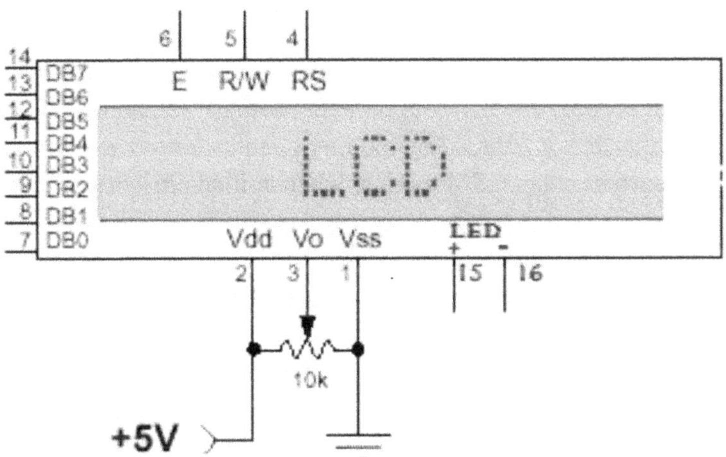

c) Result of sound alert
d) Message alert for farmer regarding animal entry

In the proposed system, Crop monitoring is done where sensors are used to collect information in the agricultural field. In our proposed work, Arduino, LDR sensor and GSM is used.

Figure 3.

When animals come close to the sensor, it detects the animal movement by generating a output signal.

After processing the signal it acts input for microcontroller which activates the system immediately buzzer will be on, at the same time it sends an SMS to the owner. The outputs of the sensor are read by the microcontroller block. SMS is generated and sent to the farmer using GSM module when animal presence is detected which alerts the farmer.

With the above stated aspects, the design has been created for easy installation of the hardware in every possible place across the globe. The objectives of the proposed project are as follows.

- We planned to design a framework to protect the agricultural fields from animals using automatic laser fence without harming animals.
- To design a security system for farm protection and prohibit the entry of animals into the farm.
- Use buzzer module for alerting the owner of the crops.
- Design a system that produces sounds in a relevant manner when particular animal tries to enter into the Farm.

2. LITERATURE SURVEY

In order to know the state of art works related to Smart Crop protection system, we have carried out a literature survey. The gist of cited papers is as given below,

A home embedded surveillance system proposed in (Nahatkar,), proposes a low-cost security system which uses small Piezoelectric Infrared sensor which is built around a microcontroller with ultra low alert power. Signal generated by PIR sensor is detected by the system, which senses the presence of individuals not at thermal equilibrium with the surrounding environment. Upon detection of any unauthorized person during any specific time interval, the system triggers to send an alarm & set up a call as well to a predefined number using GSM modem. MCU further directs the sensor signals to embedded system. The program initiates the Web camera which captures the images that can be viewed and analyzed further.

The security system in (Xie, Singh, Uang, Narayan, & Abbeel, 2013) is based on an embedded system includes GSM and sensor networks.

The PIR sensors are used for human movement detection. The system triggers an alarm detecting the presence of person in a specific interval of time and simultaneously sends the how many persons are intruder via message to the SMS through GSM Modem. When the security system is activated, the CCTV camera is activated. This highly reactive approach has low computational requirement. Therefore it is well suited for home surveillance system. This surveillance security system implemented using PIC micro controller, camera, GSM and sensors.

Corn farmland monitoring system (Gayathri et al., 2015) for monitoring the growing status of the corn (maize) plant continuously and intimate the agriculturist using wireless sensor network (WSN). But in practice, cultivator faces too much effort tin the farmland. This paper makes eases the work of the farmer in cultivated land through the usage of different kind of sensors. The two LDR sensors are interfaced with PIC16F877A microcontroller whereas its top array receives solar radiation for supply current and the bottom of the LDR array is for measuring leaf area index (LAI). The humidity sensor will compute the moisture level in the corn field, if the level decreases, then it automatically switches ON the DC motor. All the particulars of farmland are sent to the farmer through GSM and revel in the LCD screen.

Dr. Wilson, ELECTRIC FENCE presented in (Wilson, 2011) is a barrier that uses electric shocks to deter animals and people from crossing a boundary. The voltage of the shock may have effects ranging from discomfort to death. Most electric fences are used today for agricultural fencing and other forms of animal control, although they are also used to protect high-security areas such as military

The Fencing System in (Padula & Head, 2003) operates when an object or animal touches the fence, the electric circuit is completed by connecting to ground, resulting in a short electric shock, which passes through the object or animal. Solar fencing system is a popular and sought system for protecting premises for Agriculture, Security or High security applications.

The Smart design of microcontroller monitoring system for agriculture in (##NO_NAME##, 2014) different monitoring systems had been installed to improve the yield. The developed system monitored different agricultural parameters and transmitted them wirelessly to central unit via. Microcontroller. The real time values of different parameters were displayed at both transmitter and receivers side.

Multiple-Species Exclusion Fencing and Technology for Mainland Sites presented in (Day & Mac Gibbon, 2007) is exclusion fencing which is being used internationally to protect areas of high conservation value or to create' island' of protected habitat for negative fauna.

The methodology to overcome the Problem of human and animal injury and mortality due to the straying of wild animals out of the national parks and wild life sanctuaries by the use of automatic tracking and alert system is presented in (Datta et al., 2016). Automatic system and alert system has been implemented incorporating GSM and GPS technology in the form of device that would be attached to the body of an animal and would be continuously monitoring the position of the animal with respect to the GPS defined boundaries set up inside a wild life sanctuary or national park. In case an animal strays out of the GPS defined zone, an alarm system that will be installed in a human populated zone will go off, informing people about the approaching danger.

A simple and low cost approach for automatic animal detection on highways for preventing animal-vehicle collision using computer vision techniques is proposed in (Sharma & Shah, 2017). A method for finding the distance of the animal in real world units from the camera mounted vehicle also exists. The existing system is trained on more than 2200 images consisting of positive and negatives images and tested on various video clips of animals on highways with varying vehicle speed. As per the two-second rule, our proposed method can alert the driver when the vehicle speed is up to 35 km/h. An overall accuracy of almost 82.5% is achieved regarding detection using existing method.

A system to track the location of animal in the zoo or national parks is presented in (Shanmugasundaram & Pavithra, 2017). This system would include a temperature sensor and PIR sensor. The temperature sensor senses the temperature of each animal and PIR sensor senses the human presence. It continuously monitors the animals temperature if any variation in the temperature, It will be displayed on the LCD. The PIR [passive infrared sensor] sensor is used to monitor the human presence in restricted areas or near by the animal boundaries. When the human presence is detected the voice processor will give the alert to the people through the pre-recorded voice.

A method proposed in (Newlin Shebiah & Deeksha, 2017) combines the classifier into the cascade, which statistically guarantees the background extraction and pictures only the object of interest. The proposed system is tested with animal database and if the wild animals are detected, then the messages are sent through the GSM. The camera continuously records the scene and once the motion is detected, then it captures the images and makes use of the classification methods to distinguish its characteristic features. Thus, once the animals are identified as dangerous using the classification methods, the instant messages are sent to the farmers and the concerned people to safeguard themselves and their agricultural lands. The drawback of this method ids that it cannot be best suited to cover a large area of land and is not likely possible for detection and tracking of the motion of multiple or groups of animals.

Surveillance of video that deals with the monitoring of transient and persistent in a specific environment in real time. The mainly focuses on the manifest atomization of scenes and foretell the various interactions. First get input video and slit it to frames then background subtraction is done for the object and morphological features are extracted and passed to SIFT algorithm for feature extraction and then using Euclidean distance classification is done. The drawback of this system is that if the training dataset is not proper this may give wrong results.

The above literature survey study shows the work carried out by various authors who have worked on Smart Crop protection System.

2.1 Smart Crop Protection System

This framework uses a security alarm system which monitors the farm land.

This is a simple and useful security system and easy to install. Low cost and readily available components are used to realize the vibration detector. This arrangement can also be used in a rolling shutter guard for shops and shops.

The vibrations are sensed by the detector which are caused by activities such as drilling and switches on the connected load (Laser, LDR, piezo buzzer, etc) to generate an alert signal.

There are several issues with the system. It cannot distinguish the voltage drop which happens occasionally and generates an alert signal. The owner has to check manually by visiting the site where the fence is implemented. As a solution a laser based fence management is developed using wireless communication. This enables the owners to state of the fence and monitors it from remote locations safely.

The system includes several observers along with a display unit. With this arrangement the farmers will be able to measure and display voltage at the fence.

The observers transmit the condition of voltage along with the direction of the voltage leak to the display.

Here we are using 2*16 display, the display shows the received data and the farmer can know the state of the laser fence.

In his method, laser fences are used to protect the crops from the wild animals. But due to high electricity animals might get hurt which not only affecting the life of wild animals but also dangerous to the pet animals and even human beings.

One more drawback of several present existing crop protection methods is the usage of camera which is used for detection of animals which makes it economically high in cost. The alert system alerts only the forest officer and not the people near to or living in the farm which makes the response to the animal attack very slow resulting in great loss.

Traditional methods of detecting animals in farms include the use of human eyes to witness animal movements. When it is done for large farms, it is usually not possible for human beings to monitor the animal movements continuously throughout the day. Animals vary in size and nature of movement. This makes the surveillance and tracking of the wild animals difficult.

Many detection algorithms have been developed by human beings related to this. Another popular approach used by the farmers is the electrical fencing. But the drawback of this method is the huge investment which increases with the size of the farm which is to be monitored. The loss of animals lives due to electric shock is one more important aspect that needs to be taken care.

Many projects are in motion to modify this problem and a few of them are quite feasible.

Drawback: In this process only, neighbors present in the nearby vicinity of the farm will be alerted and there is no image capture of the thief to trace him later.

Animal's interference in farms will impact negatively in many ways such as depredation of crops, water supplies, damaging grain stores, injuring and death of humans.

In the proposed system, the entire process is done by using laser fencing, camera and it consists of two stages for animal detection. We have used the sensors for presence of animal in the farm vicinity. Next camera is used for detection of animal. The image is captured and image is sent for processing. Once the processing has been done the appropriate alarm (Buzzer) from the data set is sent to the farmer mobile in a very economical manner.

3. PROTOTYPE MODEL

3.1 Flow Chart

When we on the camera in the farm it starts to capture the images continuously when an animal enters the farm if not it is again sent back to the camera to capture images when an animal is not found. If image of a wild animal is captured then

Figure 4.

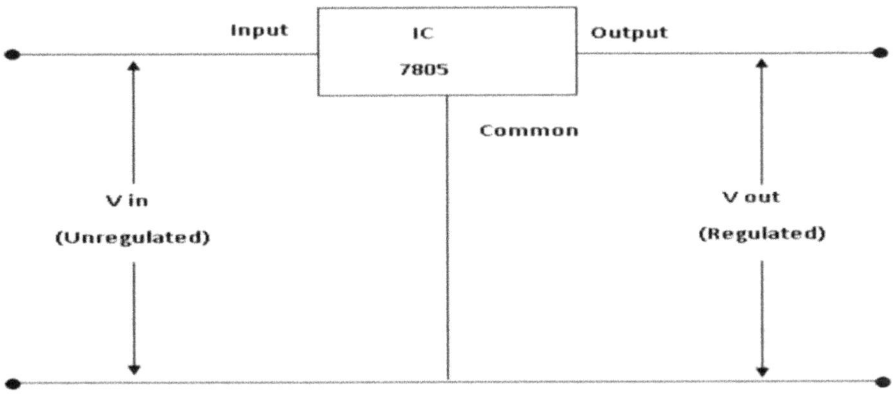

Figure 5.

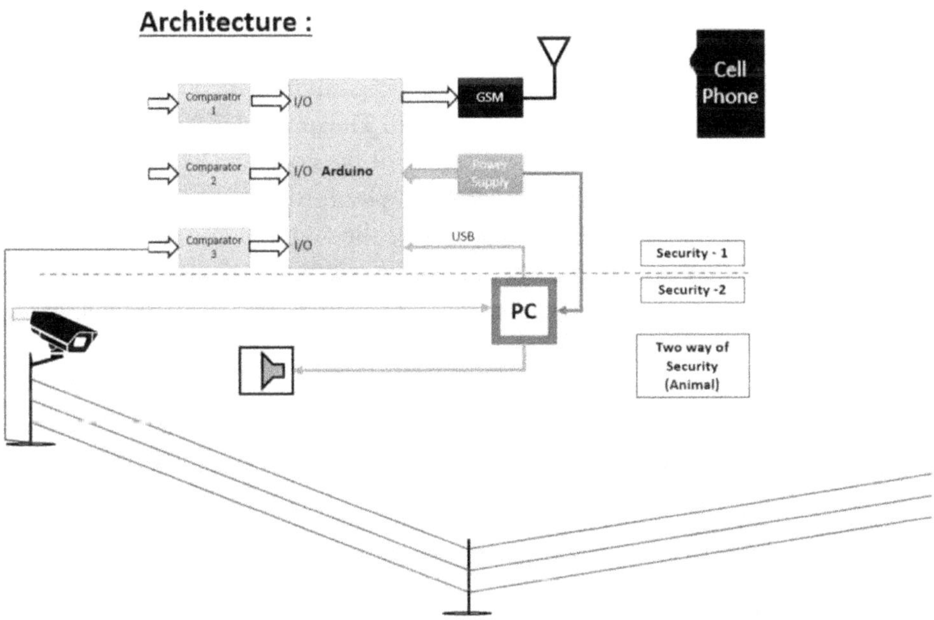

the image is sent for pre-processing and detection of animal whether it matches the data we have trained.

Animal detection is a most crucial part of the animal-human cohabitation system. Once the animal is detected the camera is triggered to capture mainly images, which then needs to be processed to find animals based on their specific features and the processed data need to be sent to the GSM further to farmer and megaphone is provided to generate the specific sound to corresponding animal for escape animals while entering into fields, then it is sent back to image capturing to capture the image.

To analyse the performance of the proposed system, we design the following: Detecting system – (i) LDR, (ii) Comparator (Fixed voltage regulator), (iii) Arduino UNO R3 (iv) GSM Module (SIM800C), (v) Farmer's mobile phone.

- If any animal crosses the laser fencing, laser light falls on animal instead of LDR.
- Comparator generates the signal of detection to i/p pin of the Arduino board.
- Arduino will result in alerting by alarm action, simultaneously generates the msg about animal detection.
- That msg will be sent to farmer's mobile phone trough the GSM module.

3.2 SOFTWARE REQUIREMENTS

3.2.1 Python

Python is predominately a dynamic typed programming language which was initiated by Guido van Rossum in the year 1989. The major design philosophy that was given more important was the readability of the code and expressing an idea in fewer lines of code rather than the verbose way of expressing things as in C++ and java. Python compiles to an intermediate code and this in turn is interpreted by the Python runtime environment to the native machine code. During its design, the creators had conceptualized the language as being a very extensible language, and hence they had designed the language to have a small core library which was extended by a huge standard library. Thus, as a result Python is used as a scripting language as it can be easily embedded into any application, though it can be used to develop a full-fledged application.

5.1.1.1 Why Python is in Trend?

- Python is open source.
- Python is also used for hardware testing.
- Google makes extensive use of python for web search system.
- Python is being applied in real revenue-generating products by real companies.

5.1.1.2 Advantages of Python

- Python is predominantly used as a scripting language for developing standalone applications that are being developed with Static-Typed language, because of the flexibility it provides due to its dynamic typed nature.
- Python favors rapid application development, which qualifies it to be used for prototyping, to a certain extent.
- Python is also used in developing websites. Due to its dynamic typing and the presence of a Virtual Machine.

5.1.2 Arduino IDE

- Arduino IDE (Integrated Development Environment) is the software for Arduino.
- It is a text editor like a notepad with different features.
- It is used for writing code, compiling the code to check if any errors are there and uploading the code to the Arduino.
- It is a cross-platform software which is available for every Operating System like Windows, Linux, macOS.
- It supports C/C++ language.
- It is open-source software, where the user can use the software as they want it to. They can also make their own modules/functions and add them to the software.
- It supports every available Arduino board including Arduino mega, Arduino Leonardo, Arduino Ethernet and more.
- Word file is called a Document similarly, Arduino file is called a Sketch where the user writes code.
- The format of Arduino is saved as .ino

5.1.2.1 How Arduino IDE Works

When a user writes code and compiles, the IDE will generate a Hex file for the code. (Hex file are Hexa Decimal files which are understood by Arduino) and then sent to the board using a USB cable. Every Arduino board is integrated with a microcontroller, the microcontroller will receive the hex file and runs as per the code written

3.3 Hardware Requirements

Different hardware devices utilized in developing the system are described in this section. The smart crop prevention system from wild animals is developed by using Arduino UNO microcontroller, PC-Laptop (i3, 4GB RAM), LDR sensor, LCD, LASER, 12V 1A Power adapter, GSM Module. When animals come near to the LDR sensor, it detects the animal movement. After getting that initial input signal, it is passed for further processing. Then it will be given to the microcontroller. After getting that initial input signal, it is passed for further processing. Then it will be given to the microcontroller. Our system will be activated, immediately buzzer will be on, at the same time it sends an SMS to the owner. Microcontroller Block is used for reading the inputs from sensor and PC. Whole process is controlled by microcontroller. The GSM module is used for sending SMS to farmer when movement of animals is detected. It alerts the farmer that some animals try to enter into the farm.

3.3.1 Arduino Uno

The Arduino UNO, is an open-source microcontroller board based on the microchip ATmega328p.

The Arduino Uno R3 is a microcontroller board based on a removable, dual-inline-package (DIP) ATmega328 AVR microcontroller. It has 20 digital input/output pins (of which 6 can be used as PWM outputs and 6 can be used as analog inputs). Programs can be loaded on to it from the easy-to-use Arduino computer program.

The Arduino has an extensive support community, which makes it a very easy way to get started working with embedded electronics. The R3 is the third, and latest, revision of the Arduino Uno.

3.3.2 Laser Module

Laser, a device that stimulates atoms or molecules to emit light at particular wavelengths and amplifies that light, typically producing a very narrow beam of radiation. The emission generally covers an extremely limited range of visible, infrared, or ultraviolet wavelengths. Many different types of lasers have been developed, with highly varied characteristics. *Laser* is an acronym for "light amplification by the stimulated emission of radiation.

Lasers deliver coherent, monochromatic, well-controlled, and precisely directed light beams. Although lasers make poor choices for general-purpose illumination, they are ideal for concentrating light in space, time, or particular wavelengths. For example, many people were first introduced to lasers by concerts in the early 1970s that incorporated laser light shows, in which moving laser beams of different colors projected changing patterns on

Planetarium domes, concert-hall ceilings, or outdoor clouds.

3.3.3 Liquid Crystal Display (LCD)

The LCD panel used in this block interfaced with micro-controller through output port. This is a 16 character × 2Line LCD module, capable of display numbers, characters, and graphics. The display contains two internal byte-wide registers, one for commands (RS=0) and the second for character to be displayed (RS=1). It also contains a user programmed Ram area (the character RAM) character that can be formed using dot matrix that can be programmed to generate any desired. Two distinguished between these areas, the hex command byte will be signify that the display RAM address 00h is chosen.

3.3.4 GSM Module

A GSM Module is a specialized type of modem which accepts a SIM card, and operates over a subscription to a mobile operator, just like a mobile phone. From the mobile operator perspective, a GSM modem looks just like a mobile phone.

When a GSM modem is connected to a computer, this allows the computer to use the GSM modem to communicate over the mobile network. While these GSM modems are most frequently used to provide mobile internet connectivity, many of them can also be used for sending and receiving SMS and MMS messages.

3.3.5 7805 IC Regulated Power Supply

A DC power supply system, which maintains constant voltage irrespective of fluctuations in the main supply or variation in the load, is known as Regulated Power supply.

The 7805 IC referred to fixed positive voltage regulator, which provides fixed voltage 5 volts. The 7805 regulator is known as fixed voltage regulator.

Fixed –Voltage regulator design has been greatly simplified by the introduction of 3-terminal regulator ICs such as the 78xx series of positive regulators and the 79xxx series of negative regulators, which incorporate features such as built-in fold back current limiting and thermal protection, etc. These ICs are available with a variety of current and output voltages ratings, as indicated by the 'xxx' suffix; current ratings are indicated by the first part of the suffix and the voltage ratings by the last two parts of the suffix. Thus, a 7805 device gives a 5V positive output at a 1a rating, and a 79L15 device gives a 15V negative output at a 100mA rating.

3-terminal regulators are very easy to use. The regulators ICs typically give about 60dB of ripple rejection, so 1V of input ripple appears as a mere 1mV of ripple on the regulated output.

3.3.6 LDR

It is basically a photocell that works on principle of photoconductivity. When passive component is basically a resistor whose resistance value decreases when the intensity of light decreases when the intensity of light decreases. LDR's are cheap and are readily available in many size and shapes.

The following methodology is used to monitor the animal entry into the farm.

1) Laser Fencing:
   - Ø If any animal crosses the laser fencing, laser light falls on animal instead of LDR.
   - Ø Comparator generates the signal of detection to i/p pin of the Arduino board.
   - Ø Arduino will result in alerting by alarm action, simultaneously generates the message about animal detection.
   - Ø That message will be sent to farmer's mobile phone trough the GSM module.
2) Image Capture:

In this proposed system, camera act as first round of security where the animal movement is detected and then picture of animal is taken by the camera and transmit the image for processing. The image is transmitted from camera to PC in the command centre where the image processing and classification of animal is done. Once the animal is found to be threat the PC will send the signal to the system to take the appropriate action.

INPUT

Along the borders of the farm the camera are pole mounted. The numbers of cameras are used relatively twice. Once an animal is been detected, it gives the signal to the camera that takes up the image of the respective scope of area in which the signal has come from. The camera will send the image for processing and classification of animal whether it is threat or not.

PROCESSING

The images that are sent by the camera are received by the PC for classification of animal. Database is created and the set of sample images are stored in it. In the collection of the images the captured image is given as query image to the processing system. A query image and images are stored in the database where the resultant is the indices corresponding to images within images that are visually similar to the query image. The image ID's output contains the indices in ranked order, from the most to the least similar match; The contents in the query image are same as that of the stored image. Consequently alarm (Buzzer) is sent to the field's owner as alert information. If the detected object is not a threat then no alarm (Buzzer) is sent. By this way false alarm can be prevented.

OUTPUT

Once the animal is classified to be a threat, necessary actions are taken. Alarm (Buzzer) will be sent to the farmer.

4. IMPLEMENTATION OF IMAGE DETECTION

Here image data set has been created using threatening object images which is taken from trails. We have taken nearly 244 object images out of which the images split into train and test images. We train the model on train images and evaluate model using the test performances. Firstly, we start from labelling the images. We label the images using Labelling tool we create a rectangular box around the object which gives co- ordinates of the object where it lies. The image data is initially stored in xml format for each image. As the number of images is more we will have a same number of xml files to avoid complexity. So, we create a CNN file which has data of all the images. Since we are working with image dataset the size of images is more so using a binary file format to store our data has an important impact on performance of the model. Binary data occupies very little space on disk, consumes less time to copy and can be read more efficiently from disk. So we convert data into record file format. The main advantage of using Hard-Disk in PC, file format is that the data can be optimized in multiple ways. This is an advantage especially for data sets that are too large to store it in memory, as the required data (e.g. a batch) is loaded from the disk as and when required and then processed. After generation of hard disk in PC we create a label map which gives a unique id for each of the category to identify. We have used a trained model of Faster CNN. The default parameters changed to our use model by using configure file. Model configure block is about configuration of a model. These configure files are used to configure parameters to the initial setting for some of the computer codes. Here, each Model Configure specifies one model to be served, including its name and the path.

After initialization of everything we train our based on our system configuration. When each training phase has begun, the loss will be reported. As the training method progress, it will begin high and get smaller and lower. We are training on model Faster CNN Inception V2 model. It started at 3 and fell down rapidly. In this we have trained nearly 5000 steps until loss is constantly less than 0.1. It took almost twelve hours to train the model. The model may train faster on powerful CPU and GPU. After completing the training process, we export an inference graph by using checkpoints that are created while training a model. This checkpoint won't contain any description of the computation defined by the model. Weights from these checkpoint files are inserted into variable operations. After finishing training from Client script, we access the api from our side by giving the test images to detect the objects in an image. These images are led to the tensor flow serving server. Tensor Flow Serving is a versatile, high performance serving system intended to produce environments different machine learning models. Tensor Flow Serving makes deploying fresh techniques or algorithms and experiments straight forward while maintaining same server design and APIs. The input images we have given are returned with bounding boxes around the object. These Bounding boxes are imaginary boxes which are around objects that are being verified for collision, objects like pedestrians on or near to the road, other signs and vehicles. Here we used threatening objects the output image is given around those threatening objects. There is a 2D coordinate system and a 3D coordinate system that are both being used. In digital image processing, the bounding box is only the coordinates of the oblong border that absolutely covers the image when placed over a page, a screen, a canvas, or other comparable bi-dimensional background. These bounding boxes are appeared based on the detection classes and detection scores. Detection classes in object detection are the craft of identifying instances of exact class, like humans, animals and many more in a video or image. To differentiate between two objects in any image or video. Detection score is to interpret the outcome, we can explore for the score and the location of each detected object. The detection score is range between 0 and 4 which gives confidence that the object was genuinely detected.

For 0- BOAR, 1- DEER, 2- ELEPHANT, 3- MONKEY these code will be sent to Arduino kit through USB for sending the name of the animals to farmer's mobile phone.

The detected score is nearer to 4, the more confident the model is. We can decide the cut-off threshold and below the threshold will be discarded. The input image fed is given as output based on detection scores where the detection score matches to a label it returns with class on top of bounding boxes.

Detection of animals:

Alarm and protection-Once the animal of the interest is matched and the object is identified, the output is finally sent to farmer mobile where an alarm is generated and the owner that an intruder has entered the field.

5. ADVANTAGES AND DISADVANTAGES

ADVANTAGES:

- Continuous monitoring can be done
- Automatic operation
- Device can be implemented in large scale
- Requires no human supervision
- Economical
- Causes no harm to animals and humans
- More Accuracy by two-way security. i.e.,1) Laser-fencing 2) Cameras
- Animals won't be effected & never dying with reference to this project.

DISADVANTAGES:

- High maintenance cost
- Continuous network supply is needed

APPLICATIONS

- Application areas of intrusion detection system are:
- Main application is to protect the farmer and wild animals.
- This can also be implemented in forest range for preventing the animal reaching the road.
- It can be used for security system for houses.
- This unit can be used in area where human's movement is minimal. It can be placed near agricultural lands to prevent the entry of wild animals from nearby forest.

6. CONCLUSION AND FUTURE SCOPE

The problem of crop vandalization by wild animals has become a major social problem in the current time and the animals, many of which are already threatened or endangered, are often killed in retaliation or to prevent future conflicts. It requires urgent attention and an effective solution. Thus this project carries a great social relevance as it aims to address this problem. With regard to this problem, hence we have made an effort to develop the system which will monitor the field using camera. So the field zone is to be monitored continuously to prevent entry of wild animals and capture image of the intruder which will be further classified using image processing that classifies efficiently with good number of accuracy and also the image of the detected animal name is displayed for a better result so that it can be used to take suitable action for crop protection against animal attacks. This work can be further extended by sending an alert in the form of a message when the animal is detected to the nearby forest office. Furthermore it can be used to reduce animal accidents. Furthermore it can be used to reduce human wildlife conflict and also animal accidents.

REFERENCES

Ahonen, T., Hadid, A., & Pietikainen, M. (2006). Pietikainen, and M.: Face description with local binary patterns:Application to face recognition. *IEEE Transactions on Pattern Analysis and Machine Intelligence*, *28*(12), 2037–2041. doi:10.1109/TPAMI.2006.244 PMID:17108377

Bherani, A., Raut, G. N., & Kale, P. (2014). Smart design of microcontroller monitoring system for agriculture. In *International Conference on Circuit, Power and Computing Technologies*. IEEE.

Datta, N., & Sarkar, S. (2016). Automatic Tracking Alarm System for Eradication of Wild Life Injury and Mortality. *IEEE Conference*. IEEE. 10.1109/ICACCAF.2016.7748972

Datta, N., Sarkar, S., Malik, A., & Punetha, D. (2016). *Automatic tracking and alarm system for eradication of wild life injury and mortality.* 2016 2nd International Conference on Advances in Computing, Communication, & Automation (ICACCA), Bareilly, India. 10.1109/ICACCAF.2016.7748972

Day, T. & Mac Gibbon, R. (2007). *Multiple – Species Exclusion Fencing and Technology for Mainland Sites.* Project Report published by National Wildlife Research Centre.

Gayathri, T., Ragul, S., & Sudharshanan, S. (2015). Corn farmland monitoring using wireless sensor network. *International Research Journal of Engineering and Technology, 2*(8).

Gogoi, M. & Philip, S. R. (2015). Protection of crops from animals using intelligent surveillance. *Journal of Applied and Fundamental Sciences.*

Nahatkar, S., Gaur, A., & Pattewa, T. (n.d.). Design of a Home Embedded Surveillance System with Piezoelectric Infrared Sensor & Ultra-Low Alert Power. *International Journal of Advanced Research in Electronics and Communication Engineering, 1*(3).

Newlin Shebiah, R., Deeksha, B., & Aparna, S. (2017). Early Warning System From Threat Of Wild Animals Using Raspberry Pi. *SSRG International Journal of Electronics and Communication Engineering*, 132-136.

Padula, R. & Head, W. (2003). *Fencing System.* Project Report published by university of Minnesota.

Shanmugasundaram, R., & Pavithra, S. (2017). IoT based animal tracking and monitoring system in Zoo. *South Asian Journal of Engineering and Technology*, *3*(2).

Sharma, S. & Shah, D. (2016). A Practical Animal Detection and Collision Avoidance System Using Computer Vision Technique. *Special Section on innovations in electrical and computer engineering education.*

Sharma, S. U., & Shah, D. J. (2017). A Practical Animal Detection and Collision Avoidance System Using Computer Vision Technique. *IEEE Access : Practical Innovations, Open Solutions*, *5*, 347–358. doi:10.1109/ACCESS.2016.2642981

Trnovszky, T., Kamencay, P., Orjesek, R., Benco, M., & Sykora, P. (2017). *Animal recognition system based on convolutional neural network.* Semantic Scholar.

Wilson. (2011). Electric Fence. *Handbook of Texas*. Project report published by the Texas State Historical Association.

Xie, Z., Singh, A., Uang, J., Narayan, K. S., & Abbeel, P. (2013). Multimodal blending for high- accuracy Instance cognition. In *IEEE/RSJ International Conference on Intelligent Robots and Systems.* IEEE. 10.1109/IROS.2013

Chapter 10
GUI-Based End-to-End Deep Learning Model for Corn Leaf Disease Classification

G. Revathy
https://orcid.org/0000-0002-0691-1687
SASTRA University (Deemed), India

J. Jeyabharathi
Kalasalingam Academy of Research and Education, India

Madonna Arieth
https://orcid.org/0000-0002-0119-4869
Sreenivasa Institute of Technology and Management Studies, India

A. Ramalingam
SRI Manakula Vinayagar Engineering College (Autonomous), India

ABSTRACT

Food security is a major problem worldwide. Ensuring that the crops produced are both safe and wholesome is crucial not only for people as the ultimate consumers of the crops, but also for farmers. Plant diseases are responsible for a significant percentage of crop losses. This alleviates the need for a fast and accurate model to discriminate and identify plants with diseases. The base chapter chosen aims to achieve the same through deep learning. The data set used in the work was obtained from Plant Village Dataset. The work customs deuce pre-trained models, EfficientNetB0 and DenseNet121, to citation the traits of the plants. The extracted traits are then fused together through concatenation to allow the model to read the more meaningful crop trait data. This also ensures that the different sets of feature data read by the two models compensate for any feature loss during extraction. It turns out that the above method gives better results associated to other models.
DOI: 10.4018/978-1-6684-9975-7.ch010

1. INTRODUCTION

Food Security indicates the availability of food at all times and the accessibility of individuals to it. According to an article published in 'Agronomy'- factors that influence crop yield and hence food security can be grouped into three categories namely technological, biological and environmental. The amount of influence of each of these factors on crop production again varies based on region to region and crop to crop. In these factors plant diseases make up a major part of the biological group with pests, weeds and insects. It is estimated by FAO that annually the loss due to pests is up to 40% of global crop production. Every year, the global economy is affected by over $220 billion due to plant diseases and a minimum of $70 billion due to invasive insects. Especially in a highly populated country like India, which is one of the largest producers of most food crops around the world, the threat to food security is even more imminent. Hence it is imperative that new and better methods to identify plant diseases are explored.

Deep Learning, a subset of machine learning, is in use across the various industries ranging from driving to medical devices. Deep learning at it's root is a neural network with layers. These layers attempt to replicate the neural functioning of the human brain to enable it to process and learn from large amounts of data. One of the key recompences of using deep learning algorithms apart from the large amounts of data it can work with is the kind of data it can work with. Deep Learning algorithms cannister progress unstructured data, for example text and images. It mechanizes feature extraction reducing the involvement of humans and also performs pattern analysis and data classification.

The traditional methods of detecting plant diseases involve visual inspection and later detailed detection in the laboratory. This process is both time-consuming and accessible to farmers from different economic backgrounds, such as B. small farmers, not accessible. This involves the development of automated and intelligent illness detection systems that deliver faster findings and are more accessible through the use of diverse technologies such as artificial intelligence, machine learning, and deep learning. Deep learning techniques gather features from pictures and then utilise these functions to conduct classification or regression as needed.

The proposed work is a unique classification model to correctly categorize corn leaves from digital images of gray spot, leaf rust, northern leaf blight and healthy leaves. This study uses two pre-trained deep convolutional neural networks (DCNNs), EfficientNetB0 and DenseNet121. Appropriate parameter selections and feature fusion techniques are used to combine the predictive capabilities of the two models and create an end-to-end classification modelOur primary contributions are as follows:

- Characteristics are retrieved from two separate models and merged.
- A appropriate amount of parameters increases classification accuracy.

- A graphical user interface (GUI) for a deep learning model for recognising and categorising maize leaf plant diseases is suggested.

2. LITERATURE SURVEY

Deep learning algorithms were recently utilised to identify and diagnose sick plants from digital photographs, automating plant disease diagnosis and aiding laypeople in recognising damaged plants. While several deep learning techniques are being utilised to identify ill plants and enhance detection rates, the primary constraint of model parameter size remains. (Amin et al., 2022).

Early sickness identification can surpass current detection methods. Deep learning approaches based on computer vision, for example, might be used to detect diseases earlier. This article examines the disease classification and detection methodologies required to identify tomato leaf diseases in depth. This essay also assesses the benefits and drawbacks of the techniques presented. Finally, using a hybrid deep learning architecture, our study proposes an early disease detection technique for detecting tomato leaf diseases(David, 2021).

Corn is a major food crop across the world. Yet, many maize diseases are difficult to identify solely on grower experience. The misdiagnosis, on the other hand, diminishes manufacturing efficiency. Deep learning and image identification technologies have become key study directions as computer technology has advanced. To detect maize leaf disease models, we propose a convolutional neural network based on data augmentation and transfer learning. The approach initially augments the input to increase the model's generalisation and accuracy before creating a transfer learning-based convolutional neural network model. The model is then trained, the convolutional neural network's training process is expedited, and the training outcomes of the test dataset feedback network are employed. The initial step was to fine-tune the GoogLeNet pre-training network and change parameters like the optimizer and learning rate. The optimisation model was then trained using transfer learning, just as the original GoogLeNet network, ResNet18, Vgg16, and Vgg19 networks, and the results were compared(Hu, 2020).

The current generation of convolutional neural networks (CNNs) has delivered great achievements in the field of image categorization. This research proposes a unique approach to constructing a plant disease detection model using leaf image classification and deep convolutional neural networks. Because of novel training approaches and methodologies, the system can be installed quickly and easily. The suggested algorithm can recognise 13 unique types of plant ailments and discriminate plant leaves from their surroundings based on healthy leaves. This method of diagnosing plant diseases was first proposed, to the best of our knowledge(Sladojevic et al., 2016).

Diseases and pests are mostly to blame for low maize yield. Reliable, quick, and early identification of maize plant diseases is critical for effective maize production. Farmers, breeders, and researchers can successfully implement suitable controlled measures to limit the consequences of illness by discovering it early. Regrettably, a lack of experience in this field, as well as the accompanying expenditures, frequently leads to misdiagnosis of maize crop diseases, which can result in severe economic losses. Several strategies for identifying plant diseases have been developed throughout the years. In recent years, computer-aided technologies, including machine learning (ML) techniques combined with plant pictures (image-based phenotyping), have become established for detecting plant diseases. Deep learning algorithms have demonstrated excellent accuracy in performing complicated cognitive tasks like humans inside machine learning approaches(Nunoo-Mensah, 2022).

Agricultural production is affected by climate changes such as humidity, rain, extreme temperatures, etc. In addition, abiotic stresses are responsible for the development of diseases and pests on crops. The production of the plants can be improved through diagnosis as well as through the timely or early detection of the exact disease. In addition, it is very difficult to carry out accurate detection and treatment with the technique used in the diagnosis of diseases and insect pests. Few researchers have looked at predicting disease and pest infestations using machine learning algorithms. Therefore, this article presents the identification of diseases in maize crops by leaf analysis at a very early stage(Agarwal, 2019).

As a consequence, the current study suggests a deep learning-based solution for monitoring plant health, which will result in an increase in the quantity and quality of plant production. The proposed enhanced DenseNet model was 98.06% accurate(Waheed, 2020).

The author of this research proposed an automated plant disease diagnosis system that uses partial least squares (PLS) regression to choose features from an extracted deep feature set(Saeed et al., 2021).

Automatic detection of plant diseases in the wild is a tough challenge in today's smart agriculture due to the diverse symptoms and mysterious causes of plant ailments. The current technique for addressing these difficulties is to build a convolutional neural network (CNN) model that extracts visual qualities and uses these aspects to identify plant disease pictures. These methods work well in lab settings with basic backdrops, but they perform poorly when processing raw images from practise fields, which contain inherent noise(Zhao et al., 2020).

Maize is a big agricultural product that, like biofuels, is critical to the food chain and many agricultural items. Additionally, households in developing countries may rely on smallholder maize farming to meet their basic needs. Sadly, maize plants are susceptible to diseases that greatly diminish agricultural production. Moreover, extreme weather and temperatures may promote the spread of disease. The growth and pervasiveness of technology has enabled the use of technological breakthroughs in a wide range of fields. Artificial intelligence-based applications, in particular, have gained traction in a variety of fields related to image, signal, and sound identification. We seek to apply deep transfer learning to categorise three maize diseases (Cercospora leaf spot, leaf rust, and northern blight) alongside healthy plants in this research. No pre-processing or explicit feature extraction were necessary when using maize leaf pictures as input and convolution models for neural networks(Fraiwan et al., 2022).

3. METHODOLOGY

The model presented visualizes the process into a three phase framework of Data preparation, Training and Evaluation.

Firstly, four subsets of the PlantVillage dataset namely

"Corn_(maize) Cercospora_leaf_spot Gray_leaf_spot"
"Corn_(maize) Common_rust_" "Corn_(maize) Northern_Leaf_Blight" "Corn_(maize) healthy" are considered.

3.1 Data Preparation

Before the dataset is supplied as input, data pre-processing is conducted on the four subsets listed above. Data pre-processing is conducted as part of the preparation step, followed by an 8:2 split of the data into training and test data. Validation data was extracted from the training data in a separate batch. The validation set comprises 20% of the instances from the training set. The model is fed the split training set to facilitate function learning. The model's performance is tested at predefined intervals during learning by entering a separate validation set. Following training, the model's performance is compared to the test subset with which the model is unfamiliar. To avoid overfitting, data expansion is conducted on the dataset using a mix of horizontal flipping, rotation, shearing, and zooming algorithms. Lastly, the photos in the subset were downsized. This concludes the data preparation process.

3.2 Training

During the training phase, the weights of the two pre-trained models are loaded, and classification is conducted concurrently on the same data set, with both models following an average pooling layer and a fully connected layer. The weights of the individual completely linked layers of each half are summed together. This is followed by three additional totally connected layers of varying diameters. At the last phase, SoftMax activates a level for classification. The cross-entropy loss function, commonly known as the logarithmic loss function, is used to assess the classification model's performance.Because EfficientNetB0 and each feature extraction half include a pre-processing phase, data augmentation for both pre-trained models is done independently due to their varied needs. This phase in EfficientNetB0 entails rescaling the pixel values to the 0 to 1 range. DenseNet121 uses ImageNet mean and standard deviation values to normalise each channel after the same rescaling step.

3.3 Evaluation

During the assessment phase, the test set was used to assess the model's performance and verify that it did not overfit. The evaluation was based on correctly classifying an input picture as late blight, rust spot, micro pitting, or healthy. At this stage we also need to assess the risk of overfitting. This step is necessary to ensure that the model does not overfit. Overfitting occurs when the model has become very accustomed to the training data and therefore cannot perform well with previously unseen test data. The model is expected to classify each image into one of the four specified classes and scored on the accuracy of the classification.

3.3.1 DENSENET121

After a sequence of procedures, the output of each layer is fed-forward to the next layer in classic neural networks. Dense Nets vary in that each level has access to the previous levels' feature maps, the gradient loss function, and the input picture. The output maps of the layer are concatenated with the input feature maps, and each layer adds to the collective information. Dense Nets are broken down into Dense

Figure 1. Workflow of the proposed model

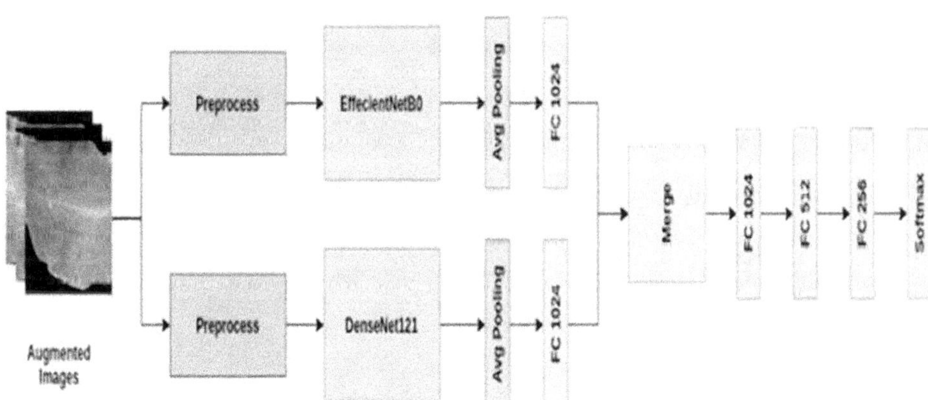

Blocks. The number of filters in these blocks varies across blocks, while the size of the feature maps remain constant inside each block. These interblock levels are known as transition layers. These layers handle down sampling by employing a stack normalisation, 1x1 convolution, and 2x2 pooling layer. DenseNet121 is a Dense Net network implementation with four dense blocks of 6,12,24,16 layers each.

3.3.2 EFFICIENTNETB0

One of the key issues in designing CNNs, is model scaling i.e providing better accuracy by increasing model size. EfficientNet is a scaling method that uses a composite coefficient to scale all dimensions of depth/width/resolution equally. EfficientNetB0 is the EfficientNet family's basic network. It was created with the maximum precision, the least amount of memory, and the highest number of floating point operations per second (FLOPS) in mind. The EfficientNet family of CNNs reduce the parameters and FLOPS to offer better accuracy and also improve efficiency. The depth of feature maps is increased by expanding input activation maps with 1x1 convolutions. The number of channels in the output feature map is reduced by subsequently performing 3x3 depth convolutions and point convolutions. The shortcut links exist between the narrow layers, while the wider layers connect the jump links.. The model size and also the number of overall operations is decreased with this structure.

3.3.3 Feature Fusion

Each convolutional layer can extract a separate set of characteristics. Numerous studies retrieved such information from multiple CNN layers and blended them to

Figure 2. Densenet architecture

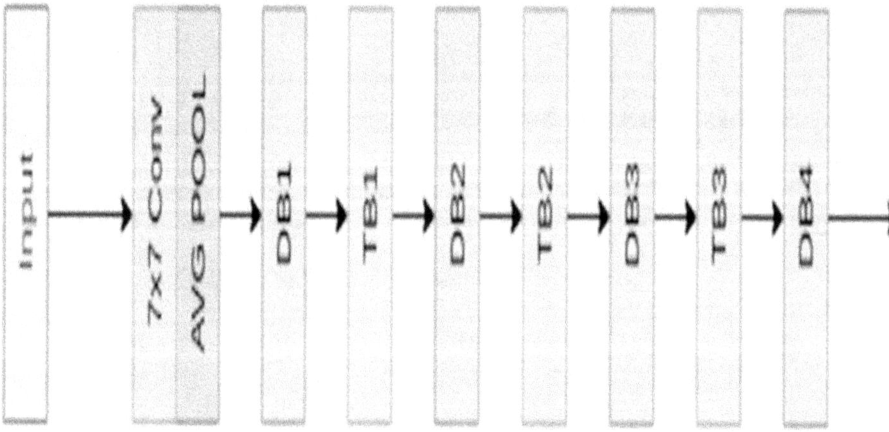

Figure 3. Efficient net architecture

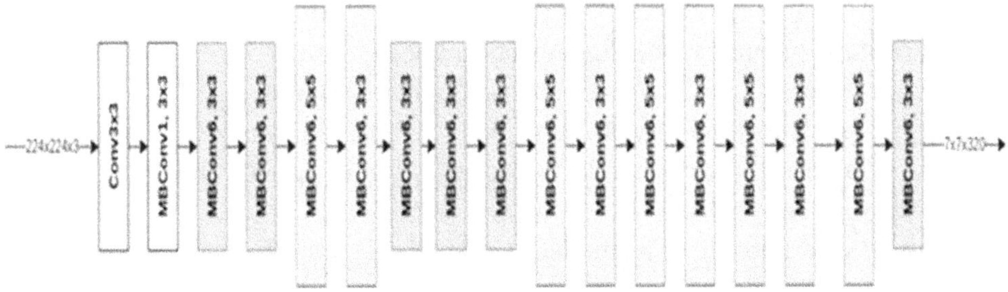

increase performance. Different features of the input data are learnt by the various CNN's leading to improved performance of the generated fused features. In layered architecture some significant features may have been lost leading to the performance improvement. Commonly feature fusion is performed in four methods namely Addition, Maximum, Average and Concatenation

In the work, concatenation method was employed to perform the fusion and improved performance was seen.

Figure 4. Feature fusion

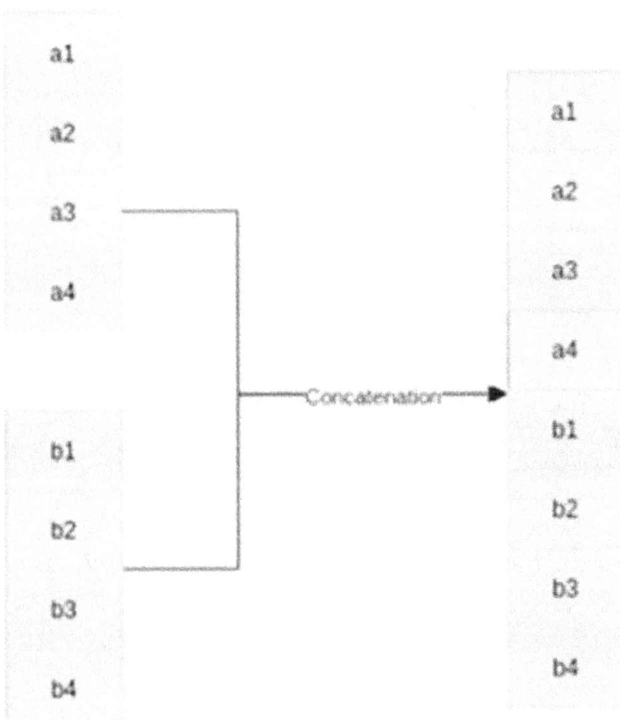

4. RESULTS AND DISCUSSIONS

The flow of the planned model, the results are been discussed.

The overall proposed flow diagram is explained in figure1.5.

Figure 6 explains the heat map generated for the initial run of the flow.

Figure 7 describes the first model Dense net 121 training and validation accuracy graphs for various epochs.

Figure 7 describes the first model Dense net 121 training and validation loss graphs for various epochs.

Figure 9 describes the second model Efficient netB0 training and validation accuracy graphs for various epochs.

Figure 10 describes the second model Efficient netB0 training and validation loss graphs for various epochs.

Figure 11 describes the final combined model accuracy which is 98.11%.

Figure 12 describes the final confusion matrix for the 4 classes we considered in the plant dataset.

The Figures 13 to 20 describes the various modes and angles of images that are taken as input and augmented.

Figure 5. Flow diagram

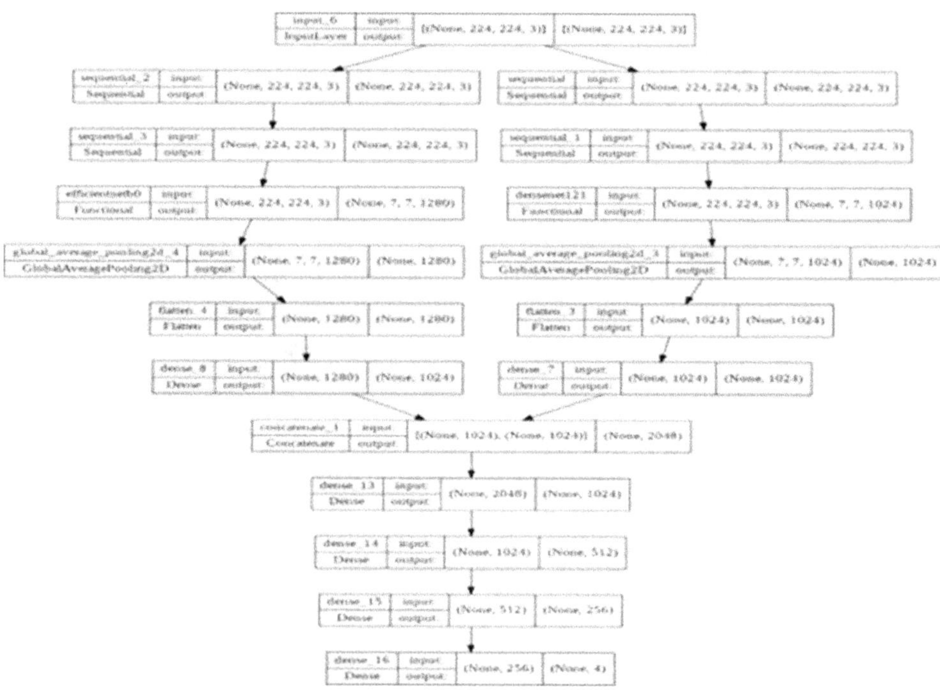

Figure 6. Heat map of initial run

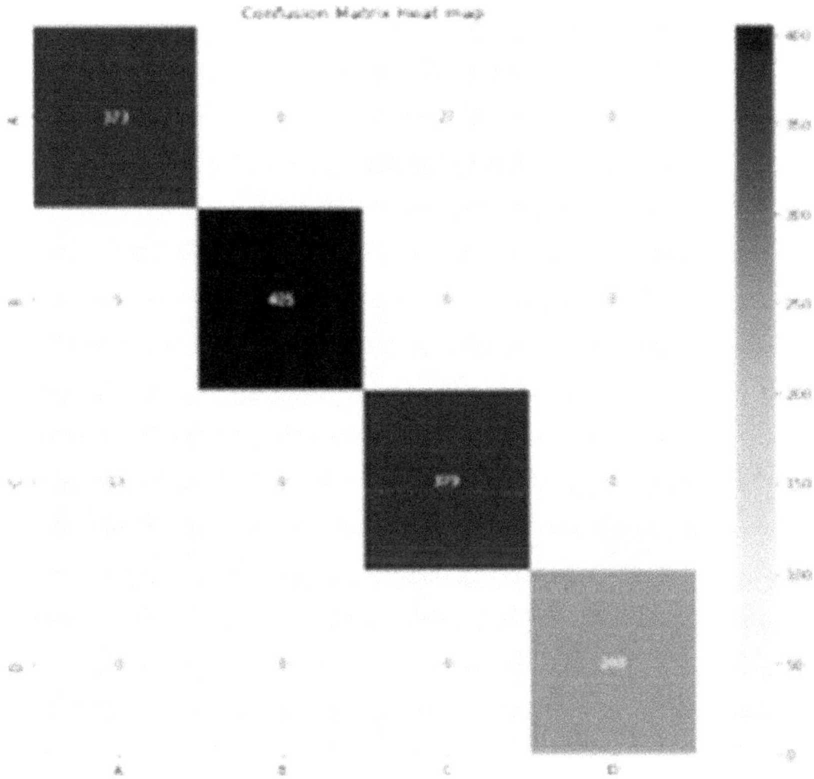

Figure 7. Densenet121 training and validation accuracy

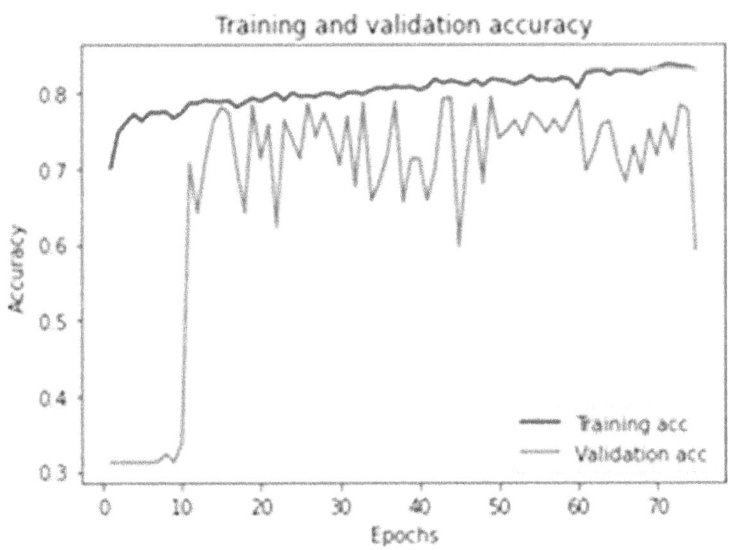

Figure 8. Densenet121 training and validation loss

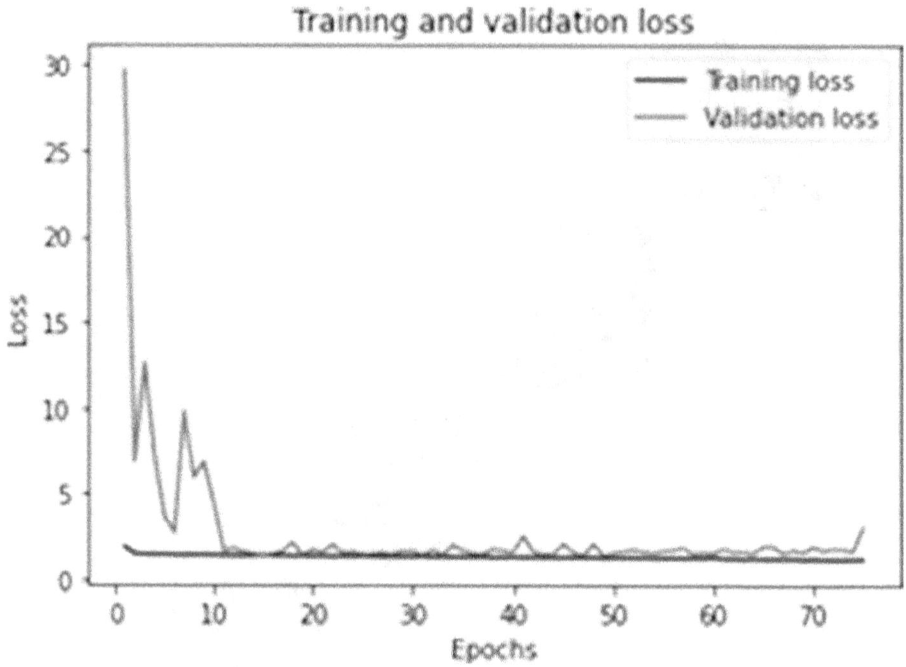

Figure 9. Efficient netB0 training and validation accuracy

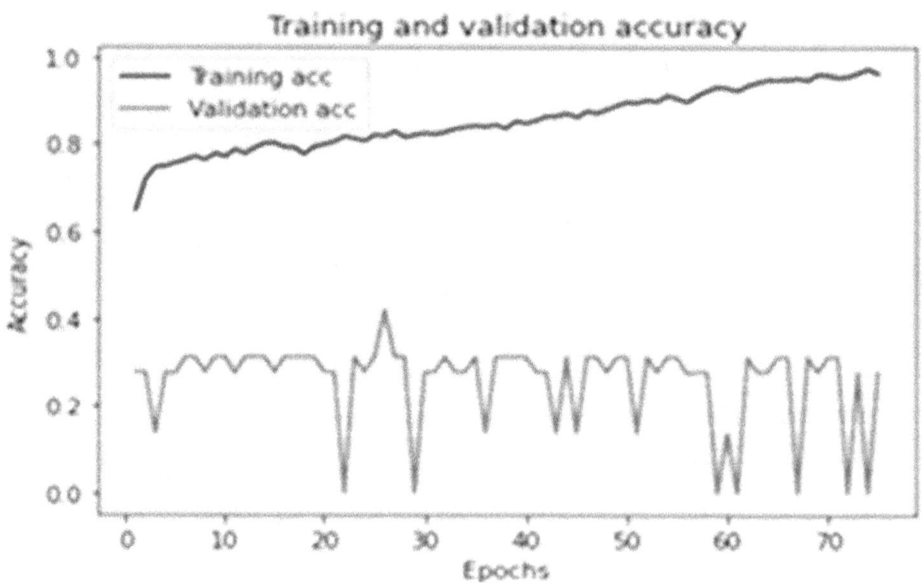

Figure 10. EfficinetnetB0 training and validation loss

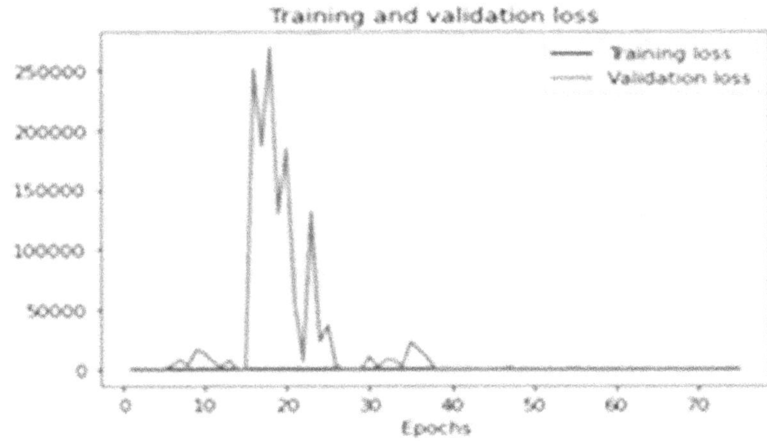

Figure 11. Final result

```
[21] scores = final_model.evaluate(test_dataset)
     print('Test loss:', scores[0])
     print('Test accuracy:', scores[1])

     97/97 [==============================] - 16s 159ms/step - loss: 0.0519 - accuracy: 0.9811
     Test loss: 0.05190702900290489
     Test accuracy: 0.981144368648529
```

Figure 12. Confusion matrix

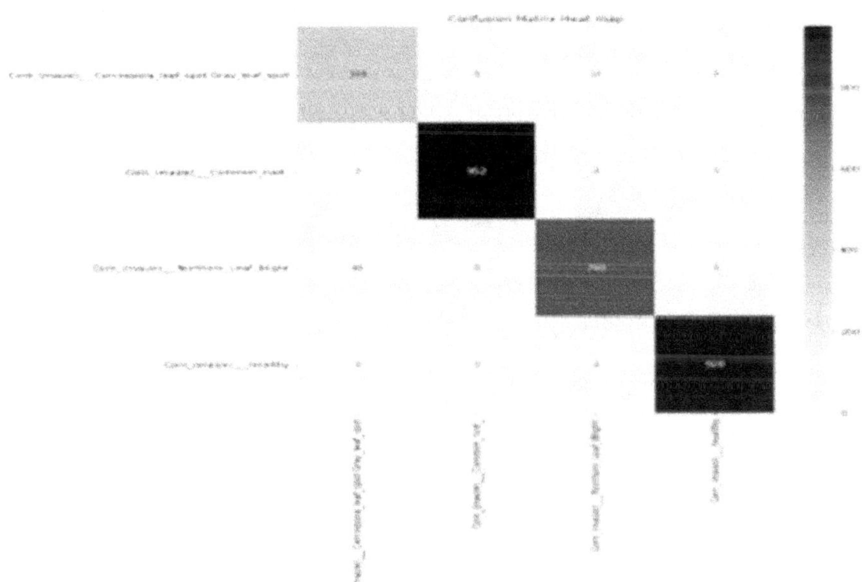

Figure 13. Random zoom in

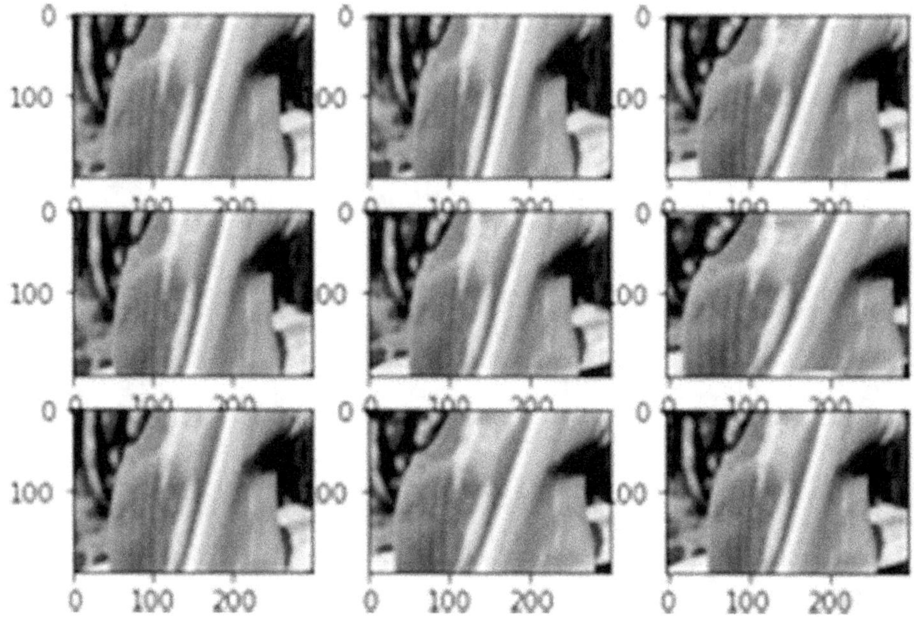

Figure 14. Random zoom out

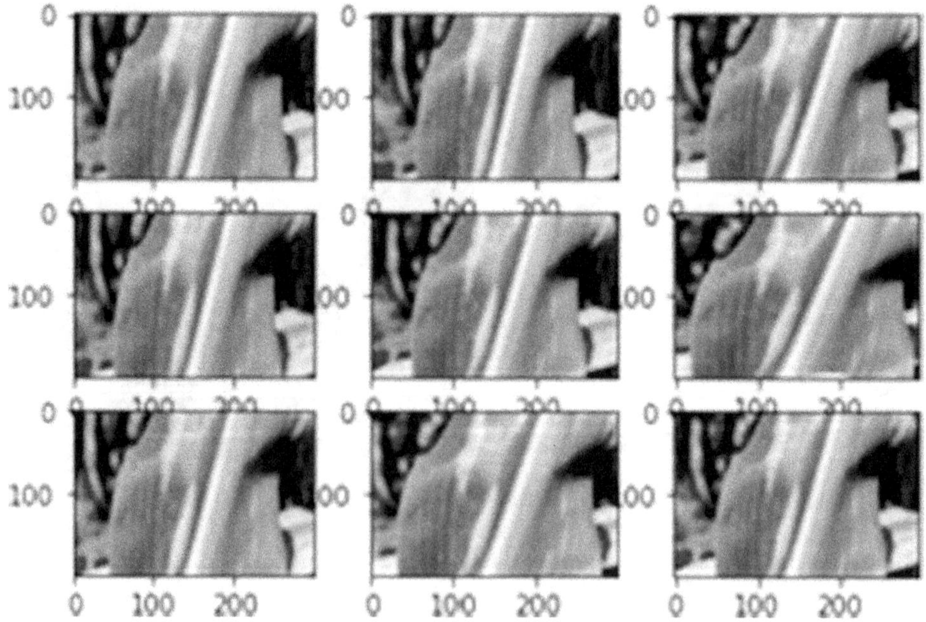

Figure 15. Random width shift

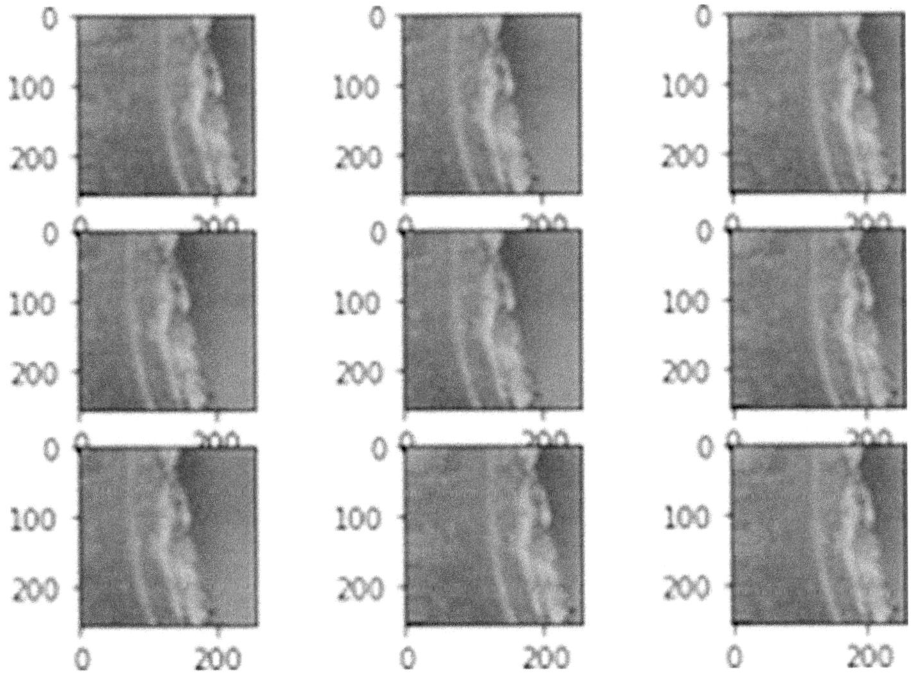

Figure 16. Random height shift

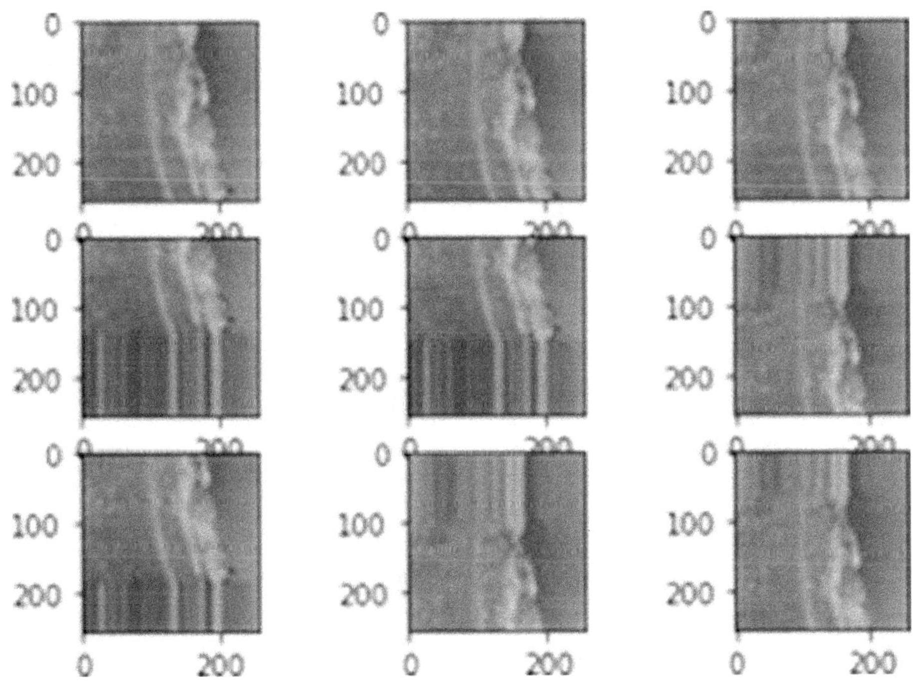

Figure 17. Random horizontal flip

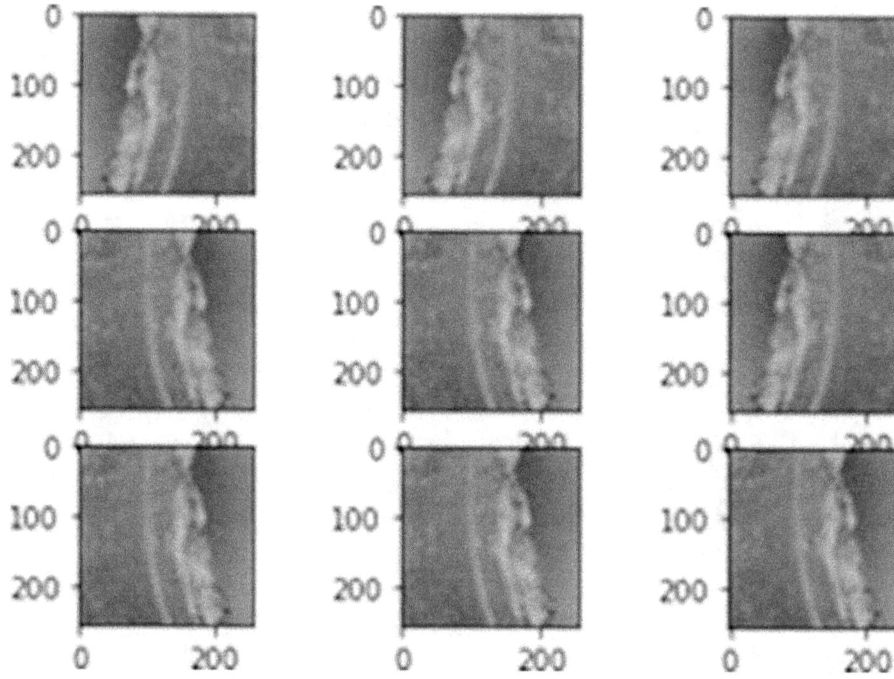

Figure 18. Random vertical flip

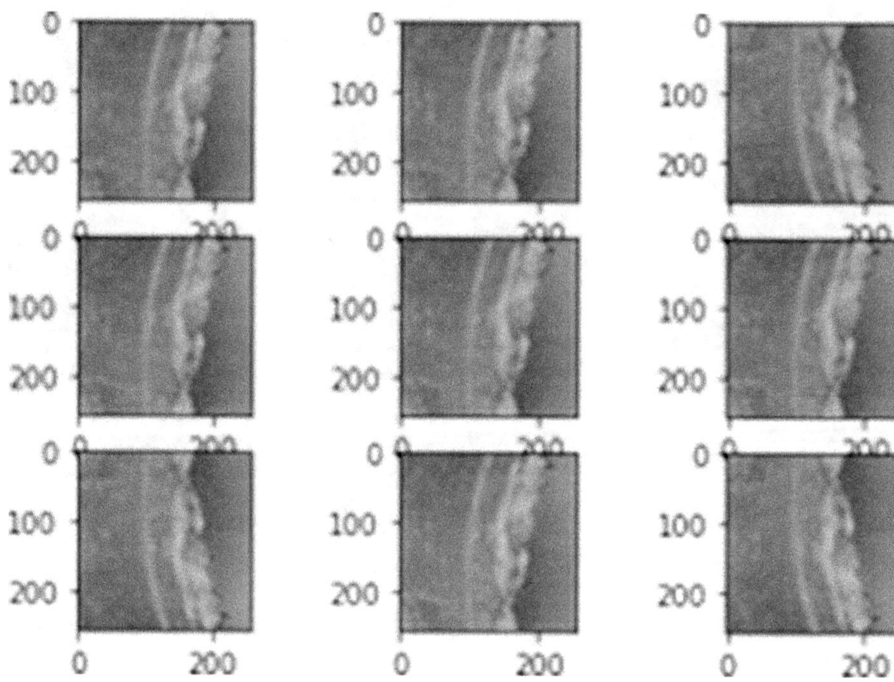

Figure 19. Random rotation

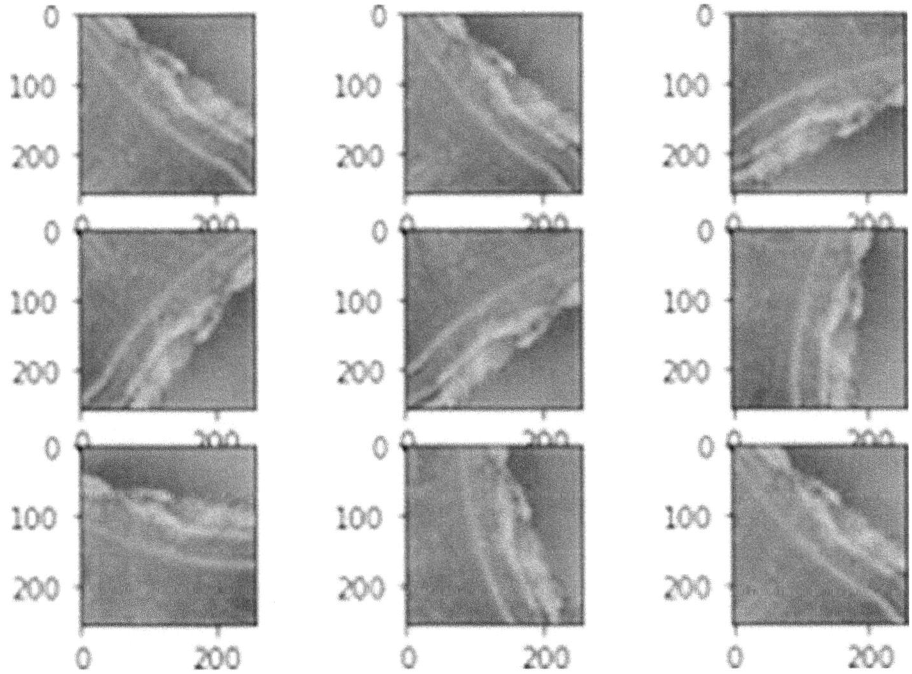

Figure 20. Random rescaling

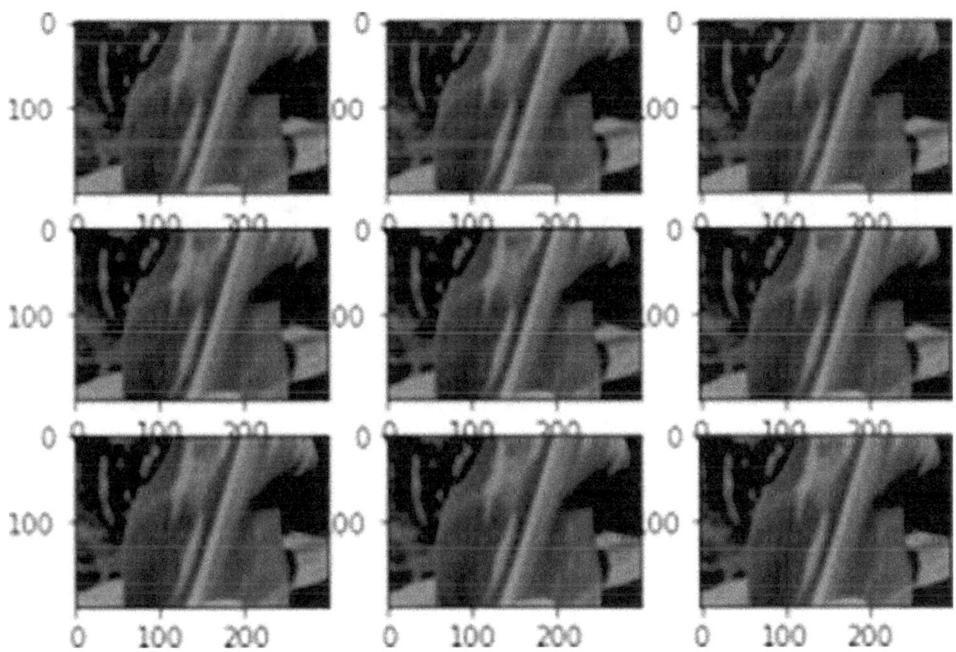

Figure 21 describes the fusion model training and validation accuracy. Figure 22 describes the fusion model training and validation loss.

Figure 21. The final model training and validation accuracy

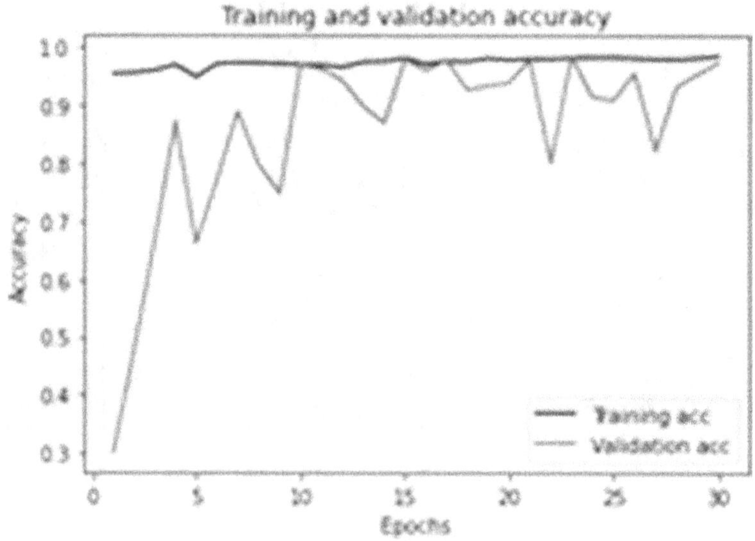

Figure 22. Training and validation loss of final model

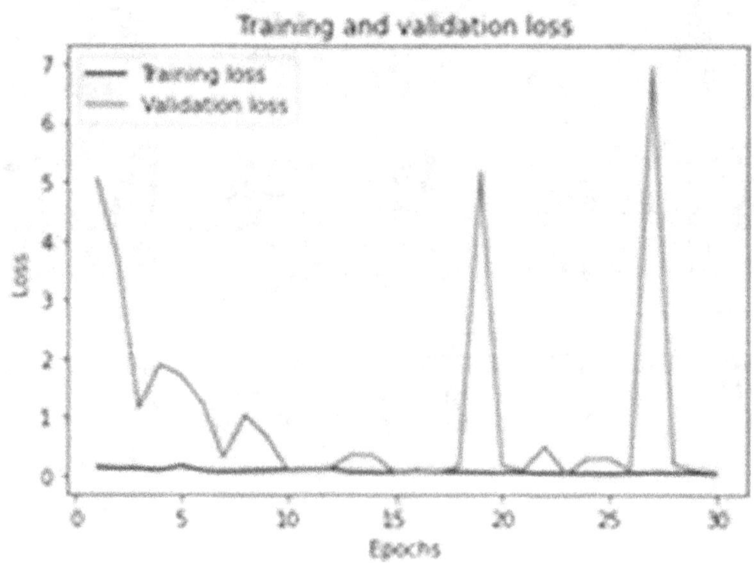

GUI APPLICATION FOR CORN LEAF DISEASES IDENTIFIER

The GUI application for the proposed model is designed by us the link is https://luissalazarsalinas-corn-leaf-diseases-corn-diseases--app-s9j6w1.streamlit.app/

The home page of our GUI gives the details about our proposed model is described in Figure 23.

Figure 24 describes how to select the corn leaf for the evaluation process in the GUI.

Figure 25 describes the selected corn leaf and the submission link for the leaf.

Figure 26 The selected Corn leaf compared with the model and the results are displayed.

Figure 27 explains the sample possibility of the 4 classes we considered for the evaluation.

Table 1 describes the final results of the confusion matrix table for the 4 classes that we considered for evaluation.

Table 2 compares the accuracy of numerous state-of-the-art models to our final model. When compared to all other state-of-the-art models, our final model has a high accuracy of 98.11%. The final model has a smaller loss than the other state-of-the-art models.

Figure 23. Homepage of the GUI

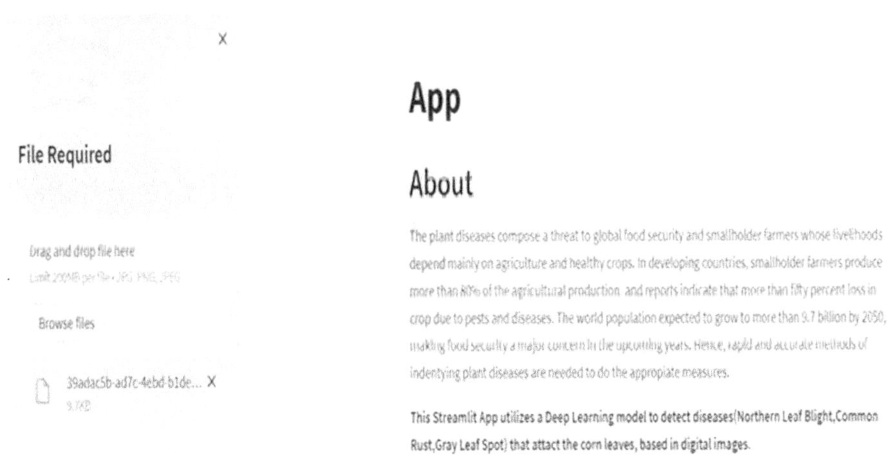

Figure 24. Selecting the corn leaf

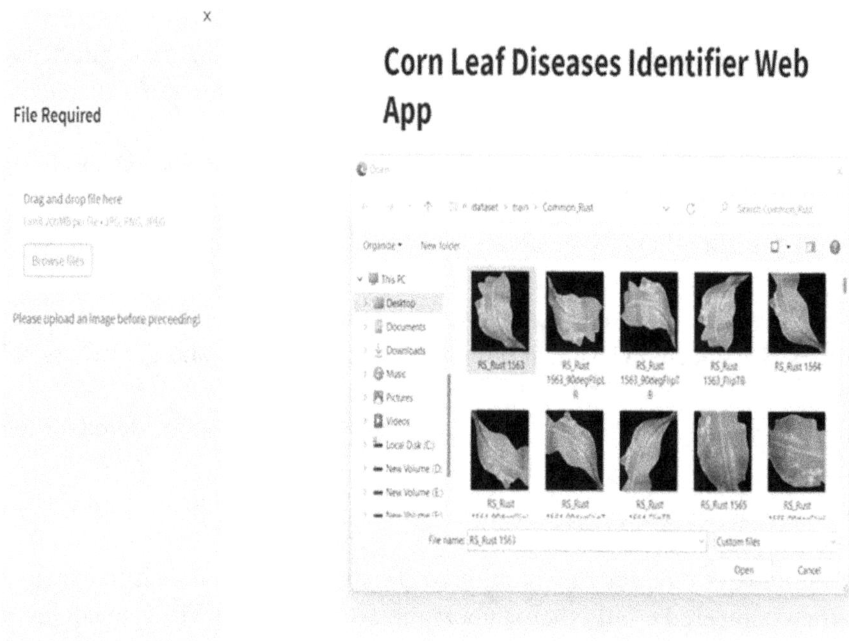

Figure 25. Selected corn leaf

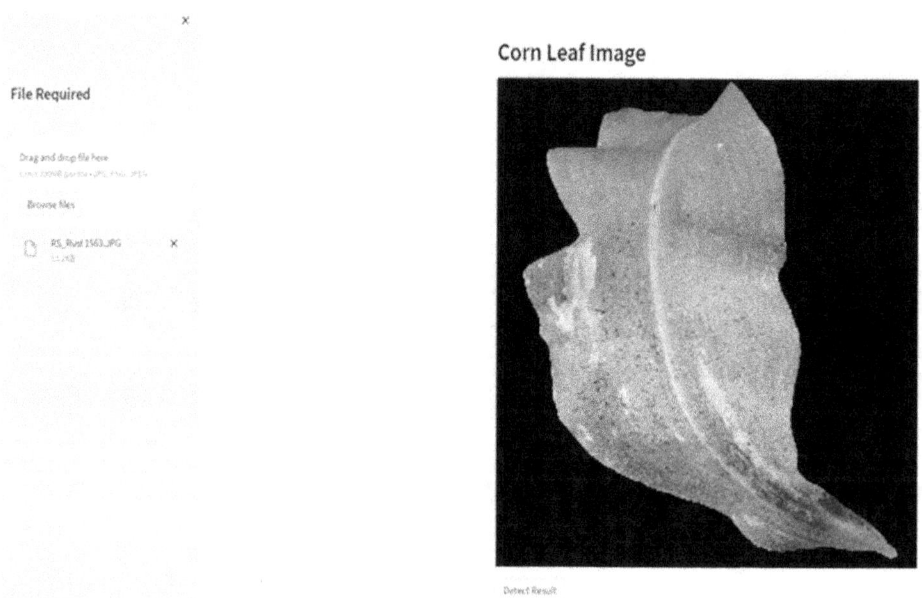

Figure 26. Description of the corn leaf

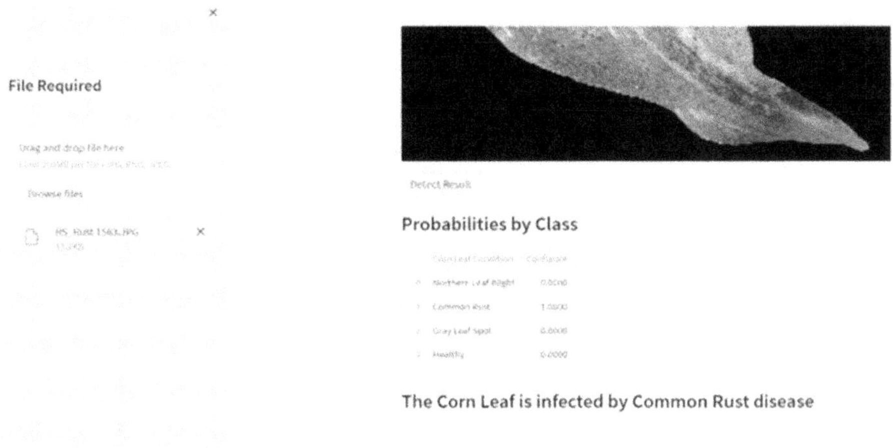

Figure 27. Sample possibility of class

Table 1. Confusion matrix percentage

Classes	Gray Leaf Spot	Common Rust	Northern Leaf Blight	Healthy
Gray leaf spot	0.975(398)	0.000(0)	0.0245(10)	0.000(0)
Common Rust	0.000(0)	1.000(952)	0.000(0)	0.000(0)
Northern Leaf Blight	0.060(48)	0.000(0)	0.940(740)	0.000(0)
Healthy	0.000(0)	0.000(0)	0.000(0)	1.000

Table 2. Accuracy table

	Accuracy	Loss
Sequential	0.9636	0.1119
Emodel	0.9763	0.0896
Dmodel	0.8307	0.1700
Final model	0.9811	0.0519

CONCLUSION

The maize dataset, a merger of four subsets of the PlantVillage dataset, was investigated. The features collected from two pre-trained models, EfficientNetB0 and DenseNet121, were combined using the feature chaining approach, and this model was employed in the implementation. Initially, the model was ran with several epoch values, and it was discovered that running the model with 30 epochs produced superior results. The suggested model showed a 98.56% accuracy. It can be demonstrated that utilizing DCNNs with lower parameters for feature extraction and then merging each set results in models that outperform typical CNN models with bigger parameters. As future study, we may apply the similar method to classify additional maize illnesses and other plant diseases from digital photos. Additional extension strategies can be investigated, as well as alternative combinations of DCNNs for feature extraction. Additionally, the findings of this study may be analyzed utilizing multiple fusion strategies for each dataset.

REFERENCES

Agarwal, M. (2019). A convolution neural network based approach to detect the disease in corn crop. *2019 IEEE 9th international conference on advanced computing (IACC)*. IEEE. 10.1109/IACC48062.2019.8971602

Amin, H., Darwish, A., Hassanien, A. E., & Soliman, M. (2022). End-to-end deep learning model for corn leaf disease classification. *IEEE Access : Practical Innovations, Open Solutions*, *10*, 31103–31115. doi:10.1109/ACCESS.2022.3159678

David, H. E. (2021). Literature Review of Disease Detection in Tomato Leaf using Deep Learning Techniques. *2021 7th International Conference on Advanced Computing and Communication Systems (ICACCS)*. IEEE. 10.1109/ICACCS51430.2021.9441714

Fraiwan, M, Faouri, E., & Khasawneh, N. (2022). Classification of Corn Diseases from Leaf Images Using Deep Transfer Learning. *Plants*, 11.

Hu, R. (2020). The identification of corn leaf diseases based on transfer learning and data augmentation. In *Proceedings of the 3rd International Conference on Computer Science and Software Engineering*. ACM. 10.1145/3403746.3403905

Nunoo-Mensah, H. (2022). A Survey of Deep Learning Techniques for Maize Leaf Disease Detection: Trends from 2016 to 2021 and Future Perspectives. *Journal of Electrical and Computer Engineering Innovations*, *10*(2), 381–392.

Saeed, F., Khan, M. A., Sharif, M., Mittal, M., Goyal, L. M., & Roy, S. (2021). Deep neural network features fusion and selection based on PLS regression with an application for crops diseases classification. *Applied Soft Computing*, *103*, 107164. doi:10.1016/j.asoc.2021.107164

Sladojevic, S., Arsenovic, M., Anderla, A., Culibrk, D., & Stefanovic, D. (2016). Deep neural networks based recognition of plant diseases by leaf image classification. *Computational Intelligence and Neuroscience*, *2016*, 2016. doi:10.1155/2016/3289801 PMID:27418923

Zhao, Y., Liu, L., Xie, C., Wang, R., Wang, F., Bu, Y., & Zhang, S. (2020). An effective automatic system deployed in agricultural Internet of Things using Multi-Context Fusion Network towards crop disease recognition in the wild. *Applied Soft Computing*, *89*, 106128. doi:10.1016/j.asoc.2020.106128

Chapter 11
A Transfer Learning Approach for Detecting Plant Leaf Diseases With Convolutional Neural Networks

P. Valarmathi
Vellore Institute of Technology, Chennai, India

N. G. Bhuvaneswari Amma
Vellore Institute of Technology, Chennai, India

Vasu Bhasin
Vellore Institute of Technology, Chennai, India

ABSTRACT

Agriculture is an indispensable sector for the continuity of homo sapiens. In the Indian context where agriculture contributes 19.9 percent of GDP and engages almost 54.6 percent of the population, it requires great measures to be taken to avoid plant diseases. Plant diseases are difficult to track manually as that requires a lot of work and proficiency in plant diseases. It is mandatory to keep a check on at varying phases of crop development to recognize the disease at time. Digitalization as well as adopting technology is essential for agriculture and for the welfare of the farmer. This chapter concentrates on detecting plant diseases using convolutional neural network. The concept of transfer learning is used for classification. The proposed framework can identify various types of diseases efficiently.

1. INTRODUCTION

About 58% of community hinge on agriculture for their livelihood and has a decisive role in the Indian economy. India ranks second in the world in terms of agriculture

DOI: 10.4018/978-1-6684-9975-7.ch011

production. Plant disease if not recognized at the correct time, may lead to huge economic issue. There are different types of diseases by various microorganisms may it be bacteria, viruses, or fungi which can damage crop to great extent and can even be fatal to human beings. Plant diseases has a detrimental impact on agriculture. Plant diseases can lead to food insecurity. Hence, it is essential to spot the disease at initial phase to bring out good crop harvest.

Plant diseases are usually identified by farmers or sometimes forestry experts but this method is inefficient. To guard them from diseases they apply pesticides beyond suitable proportion which harm ecosphere. Disease control measures are of no use and can be destructive without proper identification of the disease. One of the helpful measures in plant disease detection methods can be done with the help of image processing and neural network. Agriculture is needed sector in India as its economy depends on it. It illustrates the urgency to look after plants. The central issue is pivotal to safeguard the plant from diseases as these can affect the absolute maturation of the crop.

The proposed framework bothers with an alternate path to the advancement of disease identification, by employment of CNN. Development in machine vision gives an opportunity to strengthen plant disease identification. All crucial steps needed for executing the framework are defined all over the paper. The employment of the convolutional neural network in issue of plant disease detection has satisfactory outcomes. CNN generate ranking of visual presentation that are developed for a particular chore. CNN eradicate the need for the manual extraction for identification of disease. The CNN framework aims to look right on both ailing and healthy plants.

2. RELATED WORK

Formerly, plenty of work has been done to detect leaf diseases using machine learning and image processing. A system was proposed by Sammy et al. (2019), that was able to identify distinct varieties of apple, corn, sugarcane, grape and tomato. Hossain et al. (2019), suggested a method for recognizing the diseases with KNN. For classification, the images that were derived were used. Ch Kumari et al. (2019), proposed a method that performs computation of features like energy, coorelation mean,etc but accuracy was low. Wallelign et al. (2018) developed CNN architecture for categorization of various diseases. This framework used LeNet, for disease categorization and was able to achieve high accuracy. Jiayue et al. (2019), used YOLOv2 CNN to detect disease. YOLOv2 provides good accuracy and speed. 97% accuracy was achieved. A system was suggested by Robert G et al. (2018), using convolutional neural network for tomato leaf disease detection. This transfer learning model was able to accomplish the correctness of 95.75%. For instance,

Tran et al. (2019) in 2019 proposed a system that proctor from the seedling phase to the yielding phase. This framework was established on the idea of deep learning. This framework indicates the outcome of transfer learning on recognizing plant diseases. Konstantinos (2018) In this paper author has used a convolution neural network technique to identify various plant diseases. A detailed study has been done by the author. Images of various plant leaves are taken which includes both the infected leaves images and healthy leaves images and then the author has classified it in various classes and all CNN architectures gave more than 97% accuracy. The CNN architectures include AlexNet, AlexNetOWTBn, GoogLeNet, Overfeat, VGG. Prasanna Mohanty et al. (2016), proposed the CNN model to identify healthy and diseased plants and it was able to achieve correctness of 99.35%. The model achieves an accuracy of 80%. Parul Sharma et al. (2020). In this paper, the author has addressed a major problem with the success of a model in the real world. Most of the algorithms proposed work efficiently on training data but fail to achieve the desired output in the real-world scenario. The author made a change in training a convolutional neural network. Instead of training the convolutional neural network with a full image the author trained the model with the help of a segmented image and got very efficient output. Jitesh P. Shah et al. (2016) proposed a technique for examination of infected leaves image processing techniques and ML algorithm for identification of affected plant leaves using images. The author not only inspected, but also had briefly discussed key idea of automation for identification and detection of infected plant diseases. This proposed work was implemented to the depth where the author considered 19 experiments on various categories of diseases. This work is also utilized in the survey which was conducted for future research and upgrade of system. Shruthi et al. (2019) suggested work on predictive analytics to identify diseases. It seeks mainly data and automatically categorize the correctness of all those tasks. Proposed work demonstrate the Convolutional Neural Network (CNN) has high accuracy when a greater number of diseases are there in.

3. PROPOSED METHOD

The concept of transfer learning is used for classification. To detect plant disease, we use a convolution neural network and a deep neural network. CNNs provides systematic network for classification etc. It is simple to grasp and swift to implement. Figure 1 represents the workflow of proposed model.

Figure 1. Image dataset block diagram

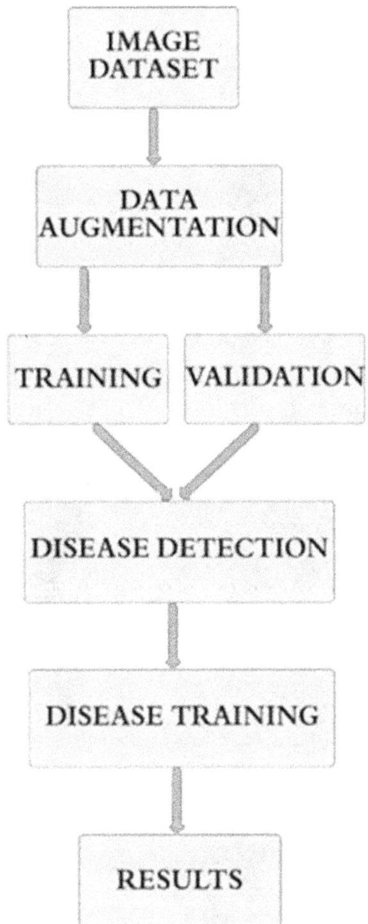

Image Acquisition

The first step is to gather data. This step involves exploring various data sources from where data can be extracted for training the model and further how the test image input is to be provided. The Plant Village Dataset is used which consist of images of plant leaves divides into 38 categories. We examine the images of plant leaves and seek to recognize the type of diseases. Figure 2 presents the images of various types of leaves.

Figure 2. Images of various types of leaves

Image Pre-Processing and Augmentation

Image augmentation is important in constructing an effective image classifier. We pre-processed the given image to improve the standard of the image. This is most important phase, as it involves the quality assurance of the data. In the image pre-processing phase image is processed to desired colour format, resized to desired size and images are denoised. To create new and different training examples data augmentation is done. Keras framework is used for image augmentation and data processing.

Image Segmentation

In this step infected part of the leaf is identified. This is again one of the most crucial step, as entire analysis is dependent on the infected region identified by the process of segmentation. It involves partitioning images into multiple segments or objects which makes identification simple.

Feature Extraction

On the basis of obtained region of interest which is the infected part of the leaf various image features like standard deviation, mean of red, blue and green channels, the entropy of image is extracted. We used colour co-occurrence method for feature extraction.

Classification of Disease

Test data has labels like early blight, healthy, and late blight, based on which classification is performed. A convolutional neural network takes an image and separate one from the other. The intention of the Convolution Neural Network is to take out the top characteristics such as colour, from the input image. Typically, the first layer is in charge to seize the bottom features. The main advantage of CNN compared to its former is that it detects the useful features without any human involvement. It grasps dissimilar characteristics by itself.

The CNN model has several layers:

1. **Input layer:** It holds information in form of picture. The specifications comprise width, height and colour (RGB format).
2. **Convolutional layer:** Convolutional layer draw out the important characteristic from the group of images.
3. **Pooling layer:** It decreases the computation power and number of parameters.
4. **Fully connected layer:** Neurons from one layer are linked to activation unit of another layer.

5. **Softmax layer:** This layer predicts the probability of the existence of an object.

Backend Model Summary

Figure 3. Backend model summary

```
dropout_2 (Dropout)           (None, 28, 28, 512)       0

conv2d_3 (Conv2D)             (None, 28, 28, 512)       2359808

batch_normalization_3 (Batc   (None, 28, 28, 512)       2048
hNormalization)

activation_3 (Activation)     (None, 28, 28, 512)       0

max_pooling2d_3 (MaxPooling   (None, 14, 14, 512)       0
2D)

dropout_3 (Dropout)           (None, 14, 14, 512)       0

flatten (Flatten)             (None, 100352)            0

dense (Dense)                 (None, 256)               25690368

batch_normalization_4 (Batc   (None, 256)               1024
hNormalization)

activation_4 (Activation)     (None, 256)               0

dropout_4 (Dropout)           (None, 256)               0

dense_1 (Dense)               (None, 512)               131584

batch_normalization_5 (Batc   (None, 512)               2048
hNormalization)

activation_5 (Activation)     (None, 512)               0

dropout_5 (Dropout)           (None, 512)               0

dense_2 (Dense)               (None, 38)                19494

=================================================================
Total params: 29,006,246
Trainable params: 29,002,278
Non-trainable params: 3,968
```

4. RESULTS AND DISCUSSIONS

After training our model on the train data with 15 epochs, the accuracy was observed to be 95.27%. Figure 3 illustrates the backend model summary. Following graphs, Figure 4 and Figure 5 show the plot of validation and training accuracy and validation and training loss. From the following plots, both loss and accuracy suppose to merge at some point.

Figure 4. Validation and training accuracy of proposed model

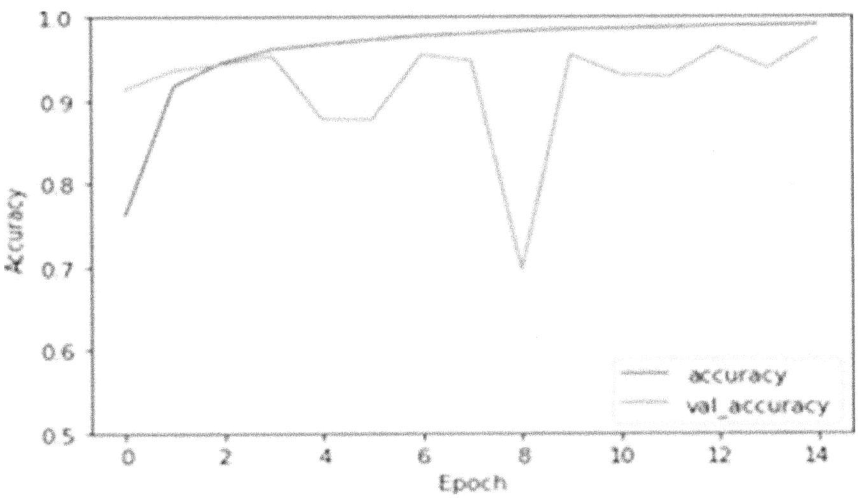

Figure 5. Validation and training loss of proposed model

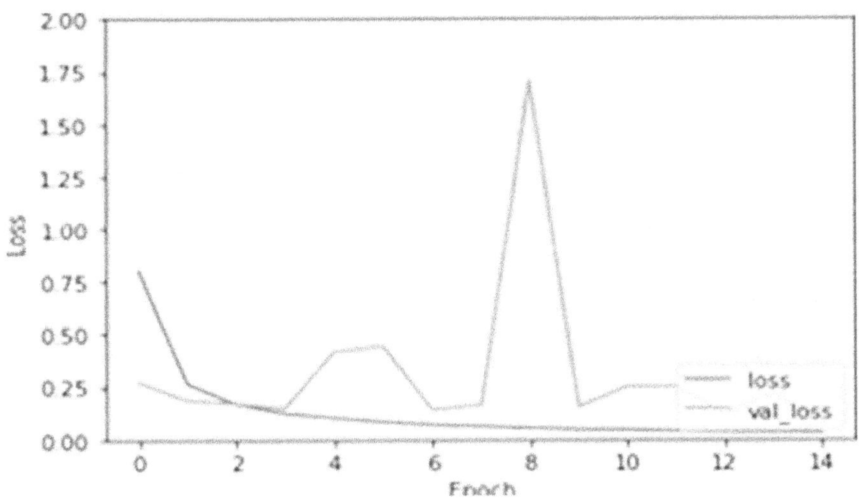

Figure 6. Prediction of disease

The proposed framework accomplished better results. Accuracy does not hold good for random guessing. The framework classified various diseases. Our model of identification of disease also helps farmers to identify the infection and do relatively controlled fertilization to avoid any future crop failures. Our work uplift farmers and it also help them to earn a better income. Prediction of disease in leaves is presented in Figure 6.

5. CONCLUSION

Plant disease identification at initial stage is difficult. Variety of machine learning techniques were employed formerly, but the accuracy was not satisfactory.

On the basis of the results observed after employing our CNN model, it can be concluded that the model can classify the plant diseases with accuracy. Moreover, the accuracy of our system can be improved by incrementing epochs. For utilization of suggested framework, it should be employed in an image-processing based software for automated determination and classification of diseases. This process should be self-regulated to show the prediction result in web application or desktop application. In addition to it agricultural department should automate the detecting the yield crops from eligibility process i.e., real time process. The most significant points in the selection of any transfer deep learning model are that the model is exercised for the allocation of diseases across the various species of crops, it utilizes the concept of deep learning and the proposed framework adopts a trending research concept of transfer learning and is able to achieve a better efficient model. If any of these points are neglected it will result in a negative transfer or overfitting problem. Nevertheless, the framework is absolutely favourable for agriculture produce and show the future of deep learning for plant disease identification.

REFERENCES

Agarwal, M., Gupta, S. K., & Biswas, K. K. (2020). Development of Efficient CNN model for Tomato crop disease identification. *Sustain. Comput. Informatics Syst.*, *28*, 100407. doi:10.1016/j.suscom.2020.100407

de Luna, R. G., Dadios, E. P., & Bandala, A. A. (2018). *Automated Image Capturing System for Deep Learning- based Tomato Plant Leaf Disease Detection and Recognition*. Proceeding of the IEEE Region 10 Conference, Jeju, South Korea.

Hossain, E., F., & Rahaman, M. (2019). *A Color and Texture Based Approach for the Detection and Classification of Plant Leaf Disease Using KNN Classifier*. Proceeding of the International Conference on Electrical, Computer and Communication Engineering (ECCE), Cox's Bazaar, Bangladesh.

Ferentinos, K. P. (2018). Deep learning models for plant disease detection and diagnosis. *Computers and Electronics in Agriculture*, *145*(January), 311–318. doi:10.1016/j.compag.2018.01.009

Militante, S. V., Gerardo, B. D., & Dionisio, N. V. (2019). Plant Leaf Detection and Disease Recognition using Deep Learning. *Proceeding of the IEEE Eurasia Conference on IOT, Communication and Engineering (ECICE)*. 10.1109/ECICE47484.2019.8942686

Mohanty, S. P., Hughes, D. P., & Salathé, M. (2016). Using deep learning for imagebased plant disease detection. *Frontiers in Plant Science*, *7*(September). PMID:27713752

Shah, J. P., Prajapati, H. B., & Dabhi, V. K. (2016). A survey on detection and classification of rice plant diseases. In *2016 IEEE Int. Conf. Curr. Trends Adv. Comput*. IEEE. 10.1109/ICCTAC.2016.7567333

Shruthi, U., Nagaveni, V., & Raghavendra, B. K. (2019). A Review on Machine Learning Classification Techniques for Plant Disease Detection. In *2019 5th Int. Conf. Adv. Comput. Commun. Syst.* (pp. 281–4). IEEE. 10.1109/ICACCS.2019.8728415

Tran, T. T., Choi, J. W., Le, T. T. H., & Kim, J. W. (2019). A comparative study of deep CNN in forecasting and classifying the macro-nutrient deficiencies on development of tomato plant. *Applied Sciences (Basel, Switzerland)*, 9.

Usha Kumari, Ch., Jeevan Prasad, S., & Mounika, G. (2019). Leaf Disease Detection: Feature Extraction with K-means clustering and Classification with ANN. *3rd International Conference on Computing Methodologies and Communication (ICCMC)*, Erode, India.

Wallelign, S., Polceanu, M., & Buche, C. (2018). Soybean plant disease identification using a convolutional neural network. In: Proc. 31st Int. Florida Artif. Intell. Res. Soc. Conf. FLAIRS, Florida, USA.

Zhao, J., & Qu, J. (2019). A Detection Method for Tomato Fruit Common Physiological Diseases Based on YOLOv2. *Proceeding of the 10th International Conference on Information Technology in Medicine and Education*. 10.1109/ITME.2019.00132

Compilation of References

Rizzo David, M., Lichtveld, M., Mazet, J. A., Togami, E., & Miller, S. A. (2011). Plant health and its effects on food safety and security in a one health framework: Four case studies. *One Health Outlook, 3*(6). PMID:33829143

Singh, D., Jackson, G., Hunter, D., Fullerton, R., Lebot, V., Taylor, M., Iosefa, T., Okpul, T., & Tyson, J. (2012). Taro leaf blight—A threat to food security. *Agriculture, 2*(3), 182–203. doi:10.3390/agriculture2030182

Rangarajan Aravind, K., & Raja, P. (2020). Automated disease classification in (Selected) agricultural crops using transfer learning. *Automatika: časopis za automatiku, mjerenje, elektroniku, računarstvo i komunikacije, 61*(2), 260-272.

Ashwinkumar, S., Rajagopal, S., Manimaran, V., & Jegajothi, B. (2022). Automated plant leaf disease detection and classification using optimal MobileNet based convolutional neural networks. *Materials Today: Proceedings, 51*, 480–487. doi:10.1016/j.matpr.2021.05.584

Borhani, Y., Khoramdel, J., & Najafi, E. (2022). A deep learning based approach for automated plant disease classification using vision transformer. *Scientific Reports, 12*(1), 11554. doi:10.103841598-022-15163-0 PMID:35798775

Singh, D., Jain, N., Jain, P., Kayal, P., Kumawat, S., & Batra, N. (2020). PlantDoc: A dataset for visual plant disease detection. In *Proceedings of the 7th ACM IKDD CoDS and 25th COMAD* (pp. 249-253). ACM. 10.1145/3371158.3371196

Fenu, G., & Malloci, F. M. (2021). DiaMOS plant: A dataset for diagnosis and monitoring plant disease. *Agronomy (Basel), 11*(11), 2107. doi:10.3390/agronomy11112107

Aswini, M., & Kavitha Bagya, S. (2022). Phytoconstituent analysis of Alpinia galanga and Zingiber officinale plant lipids–A comparative approach. *International Journal for Research Trends and Innovation, 7*(8), 647–654.

Dananjayan, S., Tang, Y., Zhuang, J., Hou, C., & Luo, S. (2022). Assessment of state-of-the-art deep learning based citrus disease detection techniques using annotated optical leaf images. *computers and Electronics in Agriculture, 193*, 106658.

ur Rahman, H., Ch, N. J., Manzoor, S., Najeeb, F., Siddique, M. Y., & Khan, R. A. (2017). A comparative analysis of machine learning approaches for plant disease identification. *advancements in life sciences, 4*(4), 120-126.

Lamba, M., Gigras, Y., & Dhull, A. (2021). Classification of plant diseases using machine and deep learning. *Open Computer Science, 11*(1), 491–508. doi:10.1515/comp-2020-0122

Jadhav, V. S. (2021). Development and advancement in image processing technique for detection of various diseases in Fruit & Vegetable Plants: A review. *Int. J. Comput. Sci. Mob. Comput.*, *10*(2), 31–38. doi:10.47760/ijcsmc.2021.v10i02.004

Vishnoi, V. K., Kumar, K., & Kumar, B. (2022). A comprehensive study of feature extraction techniques for plant leaf disease detection. *Multimedia Tools and Applications*, *81*(1), 1–53. doi:10.100711042-021-11375-0

Kurmi, Y., & Gangwar, S. (2022). A leaf image localization based algorithm for different crops disease classification. *Information Processing in Agriculture*, *9*(3), 456–474. doi:10.1016/j.inpa.2021.03.001

Shoaib, M., Hussain, T., Shah, B., Ullah, I., Shah, S. M., Ali, F., & Park, S. H. (2022). Deep learning-based segmentation and classification of leaf images for detection of tomato plant disease. *Frontiers in Plant Science*, *13*, 1031748. doi:10.3389/fpls.2022.1031748 PMID:36275583

Mohameth, F., Bingcai, C., & Sada, K. A. (2020). Plant disease detection with deep learning and feature extraction using plant village. *Journal of Computer and Communications*, *8*(6), 10–22. doi:10.4236/jcc.2020.86002

Sinshaw, N. T., Assefa, B. G., Mohapatra, S. K., & Beyene, A. M. (2022). Applications of Computer Vision on Automatic Potato Plant Disease Detection: A Systematic Literature Review. *Computational Intelligence and Neuroscience*, *2022*, 2022. doi:10.1155/2022/7186687 PMID:36419507

Filonenko, A., Kurnianggoro, L., & Jo, K. H. (2017, July). Comparative study of modern convolutional neural networks for smoke detection on image data. In *2017 10th international conference on human system interactions (HSI)* (pp. 64-68). IEEE. 10.1109/HSI.2017.8004998

Marcelino, P. (2018). Transfer learning from pre-trained models. *Towards data science, 10*, 23.

Peker, M. (2021). Multi-channel capsule network ensemble for plant disease detection. *SN Applied Sciences*, *3*(7), 707. doi:10.100742452-021-04694-2

Kinger, S., & Kulkarni, V. (2021, August). Explainable ai for deep learning based disease detection. In *2021 Thirteenth International Conference on Contemporary Computing (IC3-2021)* (pp. 209-216). 10.1145/3474124.3474154

Harakannanavar, S. S., Rudagi, J. M., Puranikmath, V. I., Siddiqua, A., & Pramodhini, R. (2022). Plant leaf disease detection using computer vision and machine learning algorithms. *Global Transitions Proceedings*, *3*(1), 305–310. doi:10.1016/j.gltp.2022.03.016

Ryo, M. (2022). Explainable artificial intelligence and interpretable machine learning for agricultural data analysis. *Artificial Intelligence in Agriculture, 6*, 257–265. doi:10.1016/j.aiia.2022.11.003

Akbar, W., Soomro, A., Ullah, M., Haq, M. I. U., Khan, S. U., & Shah, T. A. (2023, March). Performance Evaluation of Deep Learning Models for Leaf Disease Detection: A Comparative Study. *In 2023 4th International Conference on Computing, Mathematics and Engineering Technologies (iCoMET)* (pp. 01-05). IEEE. 10.1109/iCoMET57998.2023.10099223

Dahiya, S., Gulati, T., & Gupta, D. (2022). Performance analysis of deep learning architectures for plant leaves disease detection. Measurement. *Sensors (Basel), 24*, 100581.

Raschka, S. (2014). An overview of general performance metrics of binary classifier systems. arXiv preprint arXiv:1410.5330.

Arsenovic, M., Karanovic, M., Sladojevic, S., Anderla, A., & Stefanovic, D. (2019). Solving current limitations of deep learning based approaches for plant disease detection. *Symmetry, 11*(7), 939. doi:10.3390ym11070939

Buja, I., Sabella, E., Monteduro, A. G., Chiriacò, M. S., De Bellis, L., Luvisi, A., & Maruccio, G. (2021). Advances in plant disease detection and monitoring: From traditional assays to in-field diagnostics. *Sensors (Basel), 21*(6), 2129. doi:10.339021062129 PMID:33803614

Neelakantan, P. (2021). Analyzing the best machine learning algorithm for plant disease classification. *Materials Today: Proceedings*, 2214–7853.

Kartikeyan, P., & Shrivastava, G. (2021). Review on emerging trends in detection of plant diseases using image processing with machine learning. *International Journal of Computer Application, 975*, 8887.

Martinelli, F., Scalenghe, R., Davino, S., Panno, S., Scuderi, G., Ruisi, P., Villa, P., Stroppiana, D., Boschetti, M., Goulart, L. R., Davis, C. E., & Dandekar, A. M. (2015). Advanced methods of plant disease detection. A review. *Agronomy for Sustainable Development, 35*(1), 1–25. doi:10.100713593-014-0246-1

Liu, J., & Wang, X. (2021). Plant diseases and pests detection based on deep learning: A review. *Plant Methods, 17*(1), 1–18. doi:10.118613007-021-00722-9 PMID:33627131

Saleem, M. H., Potgieter, J., & Arif, K. M. (2019). Plant disease detection and classification by deep learning. *Plants, 8*(11), 468. doi:10.3390/plants8110468 PMID:31683734

Ravichandra, N. G. (2013). *Fundamentals of plant pathology*. PHI Learning Pvt. Ltd.

Rout, R., & Parida, P. (2020). A review on leaf disease detection using computer vision approach. In *Intelligent Techniques and Applications in Science and Technology: Proceedings of the First International Conference on Innovations in Modern Science and Technology 1* (pp. 863-871). Springer International Publishing. 10.1007/978-3-030-42363-6_99

Shurtleff, M. C., Pelczar, R. M., Kelman, A. & Pelczar, M. J. (2023, July 7). *plant disease*. Encyclopedia Britannica. https://www.britannica.com/science/plant-disease

Nagamani, H. S. (2019). Research Methods for Plant Health Detection using Computer Vision Techniques: A Survey. *International Journal of Applied Engineering Research: IJAER*, *14*(7), 1627–1632. doi:10.37622/IJAER/14.7.2019.1627-1632

Sokolova, M., Japkowicz, N., & Szpakowicz, S. (2006, December). Beyond accuracy, F-score and ROC: a family of discriminant measures for performance evaluation. In *Australasian joint conference on artificial intelligence* (pp. 1015-1021). Berlin, Heidelberg: Springer Berlin Heidelberg.

Abbas, F., Afzaal, H., Farooque, A. A., & Tang, S. (2020). Crop yield prediction through proximal sensing and machine learning algorithms. *Agronomy (Basel)*, *10*(7), 1046. doi:10.3390/agronomy10071046

Abdmeziem, M. R., Tandjaoui, D., & Romdhani, I. (2016). Architecting the internet of things: state of the art. *Robots and Sensor Clouds*, 55-75.

Agana, N. A., & Homaifar, A. (2017, March). A deep learning based approach for long-term drought prediction. In *SoutheastCon 2017* (pp. 1–8). IEEE. doi:10.1109/SECON.2017.7925314

Agarwal, M. (2019). A convolution neural network based approach to detect the disease in corn crop. *2019 IEEE 9th international conference on advanced computing (IACC)*. IEEE. 10.1109/IACC48062.2019.8971602

Agarwal, M., Gupta, S. K., & Biswas, K. K. (2020). Development of Efficient CNN model for Tomato crop disease identification. *Sustain. Comput. Informatics Syst.*, *28*, 100407. doi:10.1016/j.suscom.2020.100407

Agarwal, S., Agarwal, S. K., & Jain, S. K. (2014). Correlation between the stature and cranial measurements in population of North India. *ActaMedica International*, *1*(2), 99–102. doi:10.5530/ami.2014.2.10

Agarwal, S., & Tarar, S. (2021). A hybrid approach for crop yield prediction using machine learning and deep learning algorithms. *Journal of Physics: Conference Series*, *1714*(1), 012012. doi:10.1088/1742-6596/1714/1/012012

Ahearn, M. (2011). Potential challenges for beginning farmers and ranchers. In *CHOICES Online Magazine*. Agricultural & Applied Economics Association.

Ahonen, T., Hadid, A., & Pietikainen, M. (2006). Pietikainen, and M.: Face description with local binary patterns:Application to face recognition. *IEEE Transactions on Pattern Analysis and Machine Intelligence*, *28*(12), 2037–2041. doi:10.1109/TPAMI.2006.244 PMID:17108377

Akhter, R., & Sofi, S. A. (2022). Precision agriculture using IoT data analytics and machine learning. *Journal of King Saud University-Computer and Information Sciences*, *34*(8), 5602–5618. doi:10.1016/j.jksuci.2021.05.013

Alameer, A., Kyriazakis, I., Dalton, H. A., Miller, A. L., & Bacardit, J. (2020). Automatic recognition of feeding and foraging behaviour in pigs using deep learning. *Biosystems Engineering*, *197*, 91–104. doi:10.1016/j.biosystemseng.2020.06.013

Aleixandre, J. L., Aleixandre-Tudo´, J. L., Bolan˜os-Pizarro, M., & Aleixandre-Benavent, R. (2015). Mapping the scientiðc research in organic farming: A bibliometric review. *Scientometrics*, *105*(1), 295–309. doi:10.100711192-015-1677-4

Aliakbarzadeh, A., & Tabriz, A. A. (2014). Performance evaluation and ranking the branches of bank using FAHP and TOPSIS case study: Tose asr shomal interest-free loan fund. *International Journal of Academic Research in Business & Social Sciences*, *4*(12), 199. doi:10.6007/IJARBSS/v4-i12/1339

Alston, J. M., & Pardey, P. G. (2014). Agriculture in the Global Economy. *The Journal of Economic Perspectives*, *28*(1), 121–146. doi:10.1257/jep.28.1.121

Altalak, M., Uddin, M. A., Alajmi, A., & Rizg, A. (2022). A Hybrid Approach for the Detection and Classification of Tomato Leaf Diseases. *Applied Sciences (Basel, Switzerland)*, *12*(16), 8182. doi:10.3390/app12168182

Amin, H., Darwish, A., Hassanien, A. E., & Soliman, M. (2022). End-to-end deep learning model for corn leaf disease classification. *IEEE Access : Practical Innovations, Open Solutions*, *10*, 31103–31115. doi:10.1109/ACCESS.2022.3159678

Amjad, M. K., Butt, S. I., Kousar, R., Ahmad, R., Agha, M. H., Faping, Z., Anjum, N., & Asgher, U. (2018). Recent research trends in genetic algorithm based flexible job shop scheduling problems. *Mathematical Problems in Engineering*, *2018*, 2018. doi:10.1155/2018/9270802

Andrew, W., Greatwood, C., & Burghardt, T. (2019, November). Aerial animal biometrics: Individual friesian cattle recovery and visual identification via an autonomous uav with onboard deep inference. In *2019 IEEE/RSJ International Conference on Intelligent Robots and Systems (IROS)* (pp. 237-243). IEEE. 10.1109/IROS40897.2019.8968555

Antwi-Agyei, P., Dougill, A. J., & Stringer, L. C. (2015). Barriers to climate change adaptation: Evidence from northeast Ghana in the context of a systematic literature review. *Climate and Development*, *7*(4), 297–309. doi:10.1080/17565529.2014.951013

Arcury, T. A., Sandberg, J. C., Talton, J. W., Laurienti, P. J., Daniel, S. S., & Quandt, S. A. (2018). Mental Health Among Latina Farmworkers and Other Employed Latinas in North Carolina. *Rural Mental Health*, *42*(2), 89–101. doi:10.1037/rmh0000091 PMID:30237844

Ashok, T., & Varma, P. S. (2020). Crop prediction based on environmental factors using machine learning ensemble algorithms. In *Intelligent Computing and Innovation on Data Science* (pp. 581–594). Springer. doi:10.1007/978-981-15-3284-9_63

Balakrishnan, N., & Muthukumarasamy, G. (2016). Crop production-ensemble machine learning model for prediction. *International Journal of Computer Science and Software Engineering*, *5*(7), 148.

Bansod, S. & Jaiswal, R. (2018). Arduino Based Water Irrigation System. *International Journal of Advance Research in Science and Engineering, 7.*

Bao, F., & Bambil, D. (2021). Applicability of computer vision in seed identification: Deep learning, random forest, and support vector machine classification algorithms. *Acta Botanica Brasílica, 35*(1), 17–21. doi:10.1590/0102-33062020abb0361

Barbedo, J. G. A. (2018). Impact of dataset size and variety on the effectiveness of deep learning and transfer learning for plant disease classification. *Computers and Electronics in Agriculture, 153*, 46–53. doi:10.1016/j.compag.2018.08.013

Bargoti, S., & Underwood, J. (2017). Deep fruit detection in orchards. *Proceedings of the 2017 IEEE International Conference on Robotics and Automation (ICRA).*

Barhate, D. (2018). Design & Manufacturing of Multipurpose Agricultural Robot. *International Journal for Scinetific Research & Development, 6*(2).

Barman, S., Rouf, R. R., Islam, S., & Hasan, Md. M. (2017). *Proposed Design for Automated Agro-Quadcopter "Agro-ropter". 1st International Conference on Mechanical Engineering and Applied Science Military Institute of Science and Technology*, Dhaka, Bangladesh.

Behere, P. B., & Behere, A. P. (2008). Farmers' suicide in Vidarbha region of Maharashtra state: A myth or reality? *Indian Journal of Psychiatry, 50*(2), 124. doi:10.4103/0019-5545.42401 PMID:19742218

Bello, R. W., Mohamed, A. S. A., & Talib, A. Z. (2021). Contour extraction of individual cattle from an image using enhanced Mask R-CNN instance segmentation method. *IEEE Access : Practical Innovations, Open Solutions, 9*, 56984–57000. doi:10.1109/ACCESS.2021.3072636

Belton, V., & Stewart, T. (2002). *Multiple criteria decision analysis: An integrated approach.* Springer. doi:10.1007/978-1-4615-1495-4

Bhattarai, S. (2018). *New Plant Disease Dataset.* Kaggle. https://www.kaggle.com/datasets/vipoooool/new-plant-diseases-dataset

Bherani, A., Raut, G. N., & Kale, P. (2014). Smart design of microcontroller monitoring system for agriculture. In *International Conference on Circuit, Power and Computing Technologies*. IEEE.

Bian, X., Yao, G., & Shi, G. (2020). Social and natural risk factor correlation in China's fresh agricultural product supply. *PLoS One, 15*(6), e0232836. doi:10.1371/journal.pone.0232836 PMID:32484814

Bomble, P., & Lhungdim, H. (2020). Mental health status of Farmers in Maharashtra, India: A study from farmer suicide prone area of Vidarbha region. *Clinical Epidemiology and Global Health, 8*(3), 684–688. doi:10.1016/j.cegh.2020.01.002

Brigance, C., Soto Mas, F., Sanchez, V., & Handal, A. J. (2018). The mental health of the organic farmer: Psychosocial and contextual actors. *Workplace Health & Safety, 66*(12), 606–616. doi:10.1177/2165079918783211 PMID:29962329

Çakmur, H. (2014). Health Risks Faced by Turkish Agricultural Workers. *The Scientific World Journal*, *2014*, 1–8. doi:10.1155/2014/185342 PMID:25054165

Cavanagh, J. T., Carson, A. J., Sharpe, M., & Lawrie, S. M. (2003). Psychological autopsy studies of suicide: A systematic review. *Psychological Medicine*, *33*(3), 395–405. doi:10.1017/S0033291702006943 PMID:12701661

Chaharsooghi, S. K., & Ashrafi, M. (2014). Sustainable supplier performance evaluation and selection with neofuzzy TOPSIS method. *International Scholarly Research Notices*, *2014*, 2014. doi:10.1155/2014/434168 PMID:27379267

Chakraborty, S., & Prasad, K. (2018). A quality function deployment-based expert system for cotton fibre selection. *Journal of The Institution of Engineers (India): Series E*, *99*(1), 43-53.

Chamoso, P., Raveane, W., Parra, V., & González, A. (2014). UAVs applied to the counting and monitoring of animals. In *Ambient Intelligence-Software and Applications: 5th International Symposium on Ambient Intelligence* (pp. 71-80). Springer International Publishing. 10.1007/978-3-319-07596-9_8

Champaneri, M., Chachpara, D., Chandvidkar, C., & Rathod, M. (2016). Crop yield prediction using machine learning. *Technology*, *9*, 38.

Chandrakant, D. & Deshmukh, S. M. (2020). Solar Powered Multi-Function Agri-Robot. *International Research Journal of Engineering and Technology*, *7*(6).

Chang, C. L., & Lin, K. M. (2018). Smart agricultural machine with a computer vision-based weeding and variable-rate irrigation scheme. *Robotics (Basel, Switzerland)*, *7*(3), 38. doi:10.3390/robotics7030038

Chatterjee, P. (2009). Economic crisis highlights mental health issues in India. *Lancet*, *373*(9670), 1160–1161. doi:10.1016/S0140-6736(09)60666-8 PMID:19350704

Chen, S. J., & Hwang, C. L. (1992). *Fuzzy Multiple Attribute Decision Making Methods. Lecture notes in economics and mathematical system series* (Vol. 375). Springer.

Chen, Z., Wu, R., Lin, Y., Li, C., Chen, S., Yuan, Z., & Zou, X. (2022). Plant Disease Recognition Model Based on Improved YOLOv5. *Agronomy (Basel)*, *12*(2), 365. doi:10.3390/agronomy12020365

Chico, B. (2013). *Hats and Headwear around the World: A Cultural Encyclopedia*. ABC-CLIO.

Chung, D. T. P., & Van Tai, D. (2019). A fruits recognition system based on a modern deep learning technique. *Journal of Physics: Conference Series*, *1327*(1), 012050. doi:10.1088/1742-6596/1327/1/012050

Condran, S., Bewong, M., Islam, M. Z., Maphosa, L., & Zheng, L. (2022). Machine Learning in Precision Agriculture: A Survey on Trends, Applications and Evaluations Over Two Decades. *IEEE Access : Practical Innovations, Open Solutions*, *10*, 73786–73803. doi:10.1109/ACCESS.2022.3188649

Corral, S. A., de Angel, V., Salas, N., Zúñiga-Venegas, L., Gaspar, P. A., & Pancetti, F. (2017). Cognitive impairment in agricultural workers and nearby residents exposed to pesticides in the Coquimbo Region of Chile. *Neurotoxicology and Teratology*, *62*, 13–19. doi:10.1016/j.ntt.2017.05.003 PMID:28579518

Dang, V. T., Wang, J., & Dang, W. V. T. (2019). An integrated fuzzy AHP and fuzzy TOPSIS approach to assess sustainable urban development in an emerging economy. *International Journal of Environmental Research and Public Health*, *16*(16), 2902. doi:10.3390/ijerph16162902 PMID:31412685

Das, A. (2011). Farmers' suicide in India: Implications for public mental health. *The International Journal of Social Psychiatry*, *57*(1), 21–29. doi:10.1177/0020764009103645 PMID:21252353

Dash, R., Dash, D. K., & Biswal, G. C. (2021). Classification of crop based on macronutrients and weather data using machine learning techniques. *Results in Engineering*, *9*, 100203. doi:10.1016/j.rineng.2021.100203

Datta, N., & Sarkar, S. (2016). Automatic Tracking Alarm System for Eradication of Wild Life Injury and Mortality. *IEEE Conference*. IEEE. 10.1109/ICACCAF.2016.7748972

David, H. E. (2021). Literature Review of Disease Detection in Tomato Leaf using Deep Learning Techniques. *2021 7th International Conference on Advanced Computing and Communication Systems (ICACCS)*. IEEE. 10.1109/ICACCS51430.2021.9441714

Dawod, R. G., & Dobre, C. (2022). Upper and Lower Leaf Side Detection with Machine Learning Methods. *Sensors (Basel)*, *22*(7), 2696. doi:10.339022072696 PMID:35408307

Day, T. & Mac Gibbon, R. (2007). *Multiple – Species Exclusion Fencing and Technology for Mainland Sites.* Project Report published by National Wildlife Research Centre.

de Luna, R. G., Dadios, E. P., & Bandala, A. A. (2018). *Automated Image Capturing System for Deep Learning- based Tomato Plant Leaf Disease Detection and Recognition.* Proceeding of the IEEE Region 10 Conference, Jeju, South Korea.

Deccan-herald. (2020). India may become farmer suicide capitol. *Decan Harold.* https://www.deccanherald.com/national/india-may-become-farmer-suicide-capital-agri-adviser-773068.html

Dell'Acqua, F., & Gamba, P. (2003). Texture-based characterization of urban environments on satellite SAR images. *IEEE Transactions on Geoscience and Remote Sensing*, *41*(1), 153–159. doi:10.1109/TGRS.2002.807754

Department of Health and Ageing. (2013). National Mental Health Report 2013: Tracking Progress of Mental Health Reform in Australia 1993–2011. Commonwealth of Australia: Canberra, Australia.

Dhanya, V. G., Subeesh, A., Kushwaha, N. L., Vishwakarma, D. K., Kumar, T. N., Ritika, G., & Singh, A. N. (2022). *Deep learning based computer vision approaches for smart agricultural applications*. Artificial Intelligence in Agriculture. doi:10.1016/j.aiia.2022.09.007

Dharani, M. K., Thamilselvan, R., Natesan, P., Kalaivaani, P. C. D., & Santhoshkumar, S. (2021, February). Review on crop prediction using deep learning techniques. []. IOP Publishing.]. *Journal of Physics: Conference Series*, *1767*(1), 012026. doi:10.1088/1742-6596/1767/1/012026

Dierk, C., Carter, S., Chiu, P., Dunnigan, T., & Kimber, D. (2019). Use Your Head! Exploring Interaction Modalities for Hat Technologies. *Wearables*, 1033-1045. 10.1145/3322276.3322356

Din, I. U., Guizani, M., Rodrigues, J. J. P. C., Hassan, S., & Korotaev, V. V.I. U. (2019). Machine learning in the Internet of Things: Designed techniques for smart cities. *Future Generation Computer Systems*, *100*, 826–843. doi:10.1016/j.future.2019.04.017

Dongre, A. R., & Deshmukh, P. R. (2012). Farmers' suicides in the Vidarbha region of Maharashtra, India: A qualitative exploration of their causes. *Journal of Injury & Violence Research*, *4*(1), 2–7. doi:10.5249/jivr.v4i1.68 PMID:21502794

Dutta, M. (2018). If this is our land, where are our stories? Cultural piracy of crafts of Assam. *Locating Northeast India: Human Mobility, Resource Flows, and Spatial Linkages. Setting the agenda for Traditional Knowledge utilisation International Conference*, Tezpur University, India.

Eckroth, J.,& Schoen, E. (2019). *A Genetic Algorithm for Finding a Small and Diverse Set of Recent News Stories on a Given Subject: How We Generate AAAI's AI-Alert.*

Elbeltagi, A., Kushwaha, N. L., Rajput, J., Vishwakarma, D. K., Kulimushi, L. C., Kumar, M., Zhang, J., Pande, C. B., Choudhari, P., Meshram, S. G., Pandey, K., Sihag, P., Kumar, N., & Abd-Elaty, I. (2022). Modelling daily reference evapotranspiration based on stacking hybridization of ANN with meta-heuristic algorithms under diverse agro-climatic conditions. *Stochastic Environmental Research and Risk Assessment*, *36*(10), 3311–3334. doi:10.100700477-022-02196-0

Ellis, N. R., & Albrecht, G. A. (2017). Climate change threats to family farmers' sense of place and mental wellbeing: A case study from the Western Australian wheat belt. *Social Science & Medicine*, *175*, 161–168. doi:10.1016/j.socscimed.2017.01.009 PMID:28092757

Elsayed, E. A., Dawood, A. S., & Karthikeyan, R. (2017). Evaluating alternatives through the application of TOPSIS method with entropy weight. *Int. J. Eng. Trends Technol*, *46*(2), 60–66. doi:10.14445/22315381/IJETT-V46P211

Fenu, G., & Malloci, F. M. (2019, November). An application of machine learning technique in forecasting crop disease. In *Proceedings of the 3rd International Conference on Big Data Research* (pp. 76-82). 10.1145/3372454.3372474

Ferentinos, K. P. (2018). Deep learning models for plant disease detection and diagnosis. *Computers and Electronics in Agriculture*, *145*(January), 311–318. doi:10.1016/j.compag.2018.01.009

Fraiwan, M, Faouri, E., & Khasawneh, N. (2022). Classification of Corn Diseases from Leaf Images Using Deep Transfer Learning. *Plants*, 11

Fraser, C. E., Smith, K. B., Judd, F., Humphreys, J. S., Fragar, L. J., & Henderson, A. (2005). Farming and mental health problems and mental illness. *The International Journal of Social Psychiatry*, *51*(4), 340–349. doi:10.1177/0020764005060844 PMID:16400909

Ganatra, N., & Patel, A. (2021). Deep learning methods and applications for precision agriculture. Machine Learning for Predictive Analysis. *Proceedings of ICTIS, 2020*, 515–527.

Gayathri, T., Ragul, S., & Sudharshanan, S. (2015). Corn farmland monitoring using wireless sensor network. *International Research Journal of Engineering and Technology, 2*(8).

Ghadge, R., Kulkarni, J., More, P., Nene, S., & Priya, R. L. (2018). Prediction of crop yield using machine learning. *Int. Res. J. Eng. Technol., 5*.

Gogoi, M. & Philip, S. R. (2015). Protection of crops from animals using intelligent surveillance. *Journal of Applied and Fundamental Sciences*.

Golbol, F., & Olmez, H. (2021). Developing a Motion Controller for Autonomous Agricultural Robot. *2020 28th Signal Processing and Computer Applications Conference*.

Granitto, P. M., Verdes, P. F., & Ceccatto, H. A. (2005). Large-scale investigation of weed seed identification by machine vision. *Computers and Electronics in Agriculture, 47*(1), 15–24. doi:10.1016/j.compag.2004.10.003

Gregoire, A. (2002). The mental health of farmers. *Occupational Medicine (Philadelphia, Pa.), 52*(8), 471–476. PMID:12488518

Groher, T., Heitkämper, K., Walter, A., Liebisch, F., & Umstätter, C. (2020). Status quo of adoption of precision agriculture enabling technologies in Swiss plant production. *Precision Agriculture, 21*(6), 1327–1350. doi:10.100711119-020-09723-5

Grzywacz, J. G., Quandt, S. A., Chen, H., Isom, S., Kiang, L., Vallejos, Q., & Arcury, T. A. (2010). Depressive Symptoms Among Latino Farmworkers Across the Agricultural Season: Structural and Situational Influences. *Cultural Diversity & Ethnic Minority Psychology, 16*(3), 335–343. doi:10.1037/a0019722 PMID:20658876

Gupta, A., Nagda, D., Nikhare, P., & Sandbhor, A. (2021). Smart crop prediction using IoT and machine learning. *International Journal of Engineering Research & Technology (Ahmedabad)*, 2278–0181.

Gu, X., Huang, M., & Liang, X. (2019). An Improved Genetic Algorithm with Adaptive Variable Neighborhood Search for FJSP. *Algorithms, 12*(11), 243. doi:10.3390/a12110243

Habib, K. E., Gold, P. W., & Chrousos, G. P. (2001). Neuroendocrinology of stress. *Endocrinology and Metabolism Clinics of North America, 30*(3), 695–728. doi:10.1016/S0889-8529(05)70208-5 PMID:11571937

Haldurai, L., Madhubala, T., & Rajalakshmi, R. (2016). A Study on Genetic Algorithm and its Applications. *International Journal on Computer Science and Engineering, 4*, 139–143.

Hall, D., McCool, C., Dayoub, F., Sunderhauf, N., & Upcroft, B. (2015, January). *Evaluation of features for leaf classification in challenging conditions. In 2015 IEEE Winter conference on applications of computer vision*. IEEE.

Hansen, M. F., Smith, M. L., Smith, L. N., Salter, M. G., Baxter, E. M., Farish, M., & Grieve, B. (2018). Towards on-farm pig face recognition using convolutional neural networks. *Computers in Industry*, *98*, 145–152. doi:10.1016/j.compind.2018.02.016

Haralick, R. M., Shanmugam, K., & Dinstein, I. H. (1973). Textural features for image classification. *IEEE Transactions on Systems, Man, and Cybernetics*, *SMC-3*(6), 610–621. doi:10.1109/TSMC.1973.4309314

Hasan, A. M., Sohel, F., Diepeveen, D., Laga, H., & Jones, M. G. (2021). A survey of deep learning techniques for weed detection from images. *Computers and Electronics in Agriculture*, *184*, 106067. doi:10.1016/j.compag.2021.106067

Hasan, R. I., Yusuf, S. M., & Alzubaidi, L. (2020). Review of the state of the art of deep learning for plant diseases: A broad analysis and discussion. *Plants*, *9*(10), 1302. doi:10.3390/plants9101302 PMID:33019765

He, Y., & Jung, H. (2018). A voting TOPSIS approach for determining the priorities of areas damaged in disasters. *Sustainability (Basel)*, *10*(5), 1607. doi:10.3390u10051607

Hossain, E., F., & Rahaman, M. (2019). *A Color and Texture Based Approach for the Detection and Classification of Plant Leaf Disease Using KNN Classifier*. Proceeding of the International Conference on Electrical, Computer and Communication Engineering (ECCE), Cox's Bazaar, Bangladesh.

Hosseiny, B., Rastiveis, H., & Homayouni, S. (2020). An Automated Framework for Plant Detection Based on Deep Simulated Learning from Drone Imagery. *Remote Sensing (Basel)*, *12*(21), 3521. doi:10.3390/rs12213521

Huang, Y., & Jiang, W. (2018). Extension of TOPSIS Method and its Application in Investment. *Arabian Journal for Science and Engineering*, *43*(2), 693–705. doi:10.100713369-017-2736-3

Hu, R. (2020). The identification of corn leaf diseases based on transfer learning and data augmentation. In *Proceedings of the 3rd International Conference on Computer Science and Software Engineering*. ACM. 10.1145/3403746.3403905

Huzzey, J. M., Veira, D. M., Weary, D. M., & Von Keyserlingk, M. A. G. (2007). Prepartum behavior and dry matter intake identify dairy cows at risk for metritis. *Journal of Dairy Science*, *90*(7), 3220–3233. doi:10.3168/jds.2006-807 PMID:17582105

Jadav, H. R., & Shah, G. V. (2004). Determination of personal height from the length of head in Gujarat region. *Journal of the Anatomical Society of India*, *53*(1), 20–21.

Jadidi, O., Hong, T. S., Firouzi, F., Yusuff, R. M., & Zulkifli, N. (2008). TOPSIS and fuzzy multi-objective model integration for supplier selection problem. *Journal of Achievements in Materials and Manufacturing Engineering*, *31*(2), 762–769.

Jallow, M. F. A., Awadh, D. G., Albaho, M. S., Devi, V. Y., & Thomas, B. M. (2017). Pesticide Knowledge and Safety Practices among Farm Workers in Kuwait: Results of a Survey. *International Journal of Environmental Research and Public Health, 14*(4), 340. doi:10.3390/ijerph14040340 PMID:28338612

Jazi, M. S., Rezaei, A., Karami, G., Azarmi, F., & Ziejewski, M. (2014). A computational study of influence of helmet padding materials on the human brain under ballistic impacts. *Computer Methods in Biomechanics and Biomedical Engineering, 17*(12), 1368–1382. doi:10.1080/10255842.2012.748755 PMID:23281845

Jerosheja, B. R., & Mythili, C. (2020). Solar Powered Automated Multi-Tasking Agricultural Robot. *International Conference on Innovative Trends in Information Technology.*

Jiang, B., Wu, Q., Yin, X., Wu, D., Song, H., & He, D. (2019). FLYOLOv3 deep learning for key parts of dairy cow body detection. *Computers and Electronics in Agriculture, 166*, 104982. doi:10.1016/j.compag.2019.104982

Jin, X. B., Yang, N. X., Wang, X. Y., Bai, Y. T., Su, T. L., & Kong, J. L. (2020). Hybrid deep learning predictor for smart agriculture sensing based on empirical mode decomposition and gated recurrent unit group model. *Sensors (Basel), 20*(5), 1334. doi:10.339020051334 PMID:32121411

Jin, X., Che, J., & Chen, Y. (2021). Weed identification using deep learning and image processing in vegetable plantation. *IEEE Access : Practical Innovations, Open Solutions, 9*, 10940–10950. doi:10.1109/ACCESS.2021.3050296

Johnson, R. A., & Wichern, D. W. (2002). Applied multivariate statistical analysis: Vol. 5. *No. 8*. Prentice hall.

Jones-Bitton, A., Best, C., MacTavish, J., Fleming, S., & Hoy, S. (2019). Stress, anxiety, depression, and resilience in Canadian farmers. *Social Psychiatry and Psychiatric Epidemiology*. doi:10.100700127-019-01738-2 PMID:31197397

Joshi, P. P., Kanade, S. S., & Joshi, S. P. (2017). Wireless sensor network and monitoring of crop field. *IOSR Journal of Electronics and Communication Engineering, 12*(1), 23-28.

Kaggle. (2023). *Crop Production Dataset*. Kaggle. https://www.kaggle.com/datasets/atharvaingle/crop-recommendation-dataset

Kalimuthu, M., Vaishnavi, P., & Kishore, M. (2020, August). Crop prediction using machine learning. In *2020 third international conference on smart systems and inventive technology (ICSSIT)* (pp. 926-932). IEEE. 10.1109/ICSSIT48917.2020.9214190

Kallioniemi, M. K., Simola, A., Kaseva, J., & Kymäläinen, H.-R. (2016). Stress and Burnout Among Finnish Dairy Farmers. *Journal of Agromedicine, 21*(3), 259–268. doi:10.1080/1059924X.2016.1178611 PMID:27081893

Kamilaris, A., & Prenafeta-Boldú, F. X. (2018). A review of the use of convolutional neural networks in agriculture. *Journal of Agricultural Science, 156*(3), 312–322. doi:10.1017/S0021859618000436

Kamilaris, A., & Prenafeta-Boldú, F. X. (2018). Deep learning in agriculture: A survey. *Computers and Electronics in Agriculture*, *147*, 70–90. doi:10.1016/j.compag.2018.02.016

Kamyshova, G., Osipov, A., Gataullin, S., Korchagin, S., Ignar, S., Gataullin, T., Terekhova, N., & Suvorov, S. (2022). Artificial neural networks and computer vision's-based phytoindication systems for variable rate irrigation improving. *IEEE Access : Practical Innovations, Open Solutions*, *10*, 8577–8589. doi:10.1109/ACCESS.2022.3143524

Kannur, A., Kannur, A., & Rajpurohit, V. S. (2011). Classification and grading of bulk seeds using artificial neural network. *Jisuanji Yingyong*, *3*, 62–73.

Kasinathan, T., Singaraju, D., & Uyyala, S. R. (2021). Insect classification and detection in field crops using modern machine learning techniques. *Information Processing in Agriculture*, *8*(3), 446–457. doi:10.1016/j.inpa.2020.09.006

Katarzyna, R., & Paweł, M. (2019). A vision-based method utilizing deep convolutional neural networks for fruit variety classification in uncertainty conditions of retail sales. *Applied Sciences (Basel, Switzerland)*, *9*(19), 3971. doi:10.3390/app9193971

Kearney, G. D., Rafferty, A. P., Hendricks, L. R., Allen, D. L., & Tutor-Marcom, R. (2014). A cross sectional study of stressors among farmers in Eastern North Carolina. *North Carolina Medical Journal*, *75*(6), 384–392. doi:10.18043/ncm.75.6.384 PMID:25402689

Kestur, R., Meduri, A., & Narasipura, O. (2019). MangoNet: A deep semantic segmentation architecture for a method to detect and count mangoes in an open orchard. *Engineering Applications of Artificial Intelligence*, *77*, 59–69. doi:10.1016/j.engappai.2018.09.011

Khalaf, M., Alaskar, H., Hussain, A. J., Baker, T., Maamar, Z., Buyya, R., Liatsis, P., Khan, W., Tawfik, H., & Al-Jumeily, D. (2020). IoT-enabled flood severity prediction via ensemble machine learning models. *IEEE Access : Practical Innovations, Open Solutions*, *8*, 70375–70386. doi:10.1109/ACCESS.2020.2986090

Khalaf, M., Hussain, A. J., Al-Jumeily, D., Baker, T., Keight, R., Lisboa, P., ... Al Kafri, A. S. (2018). A data science methodology based on machine learning algorithms for flood severity prediction. In *2018 IEEE Congress on Evolutionary Computation (CEC)* (pp. 1–8). IEEE 10.1109/CEC.2018.8477904

Khan, I., Shinde, A., Khandelwal, S., Sah, V., & Narendiranath, B. T. (2019). FINITE ELEMENT Analysis On Different Materials For Motorcycle Helmet Shell. *International Research Journal of Engineering and Technology*, *6*(5), 4581–4586.

Khan, S., Tufail, M., Khan, M. T., Khan, Z. A., & Anwar, S. (2021). Deep learning based identification system of weeds and crops in strawberry and pea fields for a precision agriculture sprayer. *Precision Agriculture*, *22*(6), 1711–1727. doi:10.100711119-021-09808-9

Kim, Y. H., Yoo, S. J., Gu, Y. H., Lim, J. H., Han, D., & Baik, S. W. (2014). Crop pests prediction method using regression and machine learning technology: Survey. *IERI Procedia*, *6*, 52–56. doi:10.1016/j.ieri.2014.03.009

Kishore, K. V. K., Kumar, B. Y., & Venkatramaphanikumar, S. (2020). Optimized water scheduling using IoT sensor data in smart farming. In *Smart Technologies in Data Science and Communication* (pp. 205–219). Springer. doi:10.1007/978-981-15-2407-3_26

Koech, R., & Langat, P. (2018). Improving irrigation water use efficiency: A review of advances, challenges and opportunities in the Australian context. *Water (Basel)*, *10*(12), 1771. doi:10.3390/w10121771

Koh, S.-B., Kim, T. H., Min, S., Lee, K., Kang, D. R., & Choi, J. R. (2017). Exposure to pesticide as a risk factor for depression: A population-based longitudinal study in Korea. *Neurotoxicology*, *62*, 181–185. doi:10.1016/j.neuro.2017.07.005 PMID:28720389

Kolstrup, C., Lundqvist, P., & Pinzke, S. (2008). Psychosocial Work Environment Among Employed Swedish Dairy and Pig Farmworkers. *Journal of Agromedicine*, *13*(1), 23–36. doi:10.1080/10599240801986157 PMID:19042690

Kou, G., Miettinen, K., & Shi, Y. (2011). Multiple criteria decision making: Challenges and advancements. *Journal of Multi-Criteria Decision Analysis*, *18*(1-2), 1–4. doi:10.1002/mcda.476

Krizhevsky, A., Sutskever, I., & Hinton, G. E. (2017). Imagenet classification with deep convolutional neural networks. *Communications of the ACM*, *60*(6), 84–90. doi:10.1145/3065386

Kumar, R., Singh, M. P., Kumar, P., & Singh, J. P. (2015, May). Crop Selection Method to maximize crop yield rate using machine learning technique. In 2015 international conference on smart technologies and management for computing, communication, controls, energy and materials (ICSTM) (pp. 138-145). IEEE. doi:10.1109/ICSTM.2015.7225403

Kumar, S., Singh, S. K., Singh, R., Singh, A. K., Kumar, S., Singh, S. K., & Singh, A. K. (2017). Recognition of cattle using face images. *Animal Biometrics: Techniques and Applications*, 79-110.

Kumar, Y. J. N., Spandana, V., Vaishnavi, V. S., Neha, K., & Devi, V. G. R. R. (2020, June). Supervised machine learning approach for crop yield prediction in agriculture sector. In *2020 5th International Conference on Communication and Electronics Systems (ICCES)* (pp. 736-741). IEEE. 10.1109/ICCES48766.2020.9137868

Kurdi, M. M., & Elzein, I. A. (2022). Trajectory and Motion for Agricultural Robot. *2022 International Conference on Decision Aid Sciences and Applications*. 10.1109/DASA54658.2022.9765282

Kushwaha, N. L., Bhardwaj, A., & Verma, V. K. (2016). Hydrologic response of Takarla-Ballowal watershed in Shivalik foot-hills based on morphometric analysis using remote sensing and GIS. *J Indian Water Resour Soc*, *36*(1), 17–25.

Kushwaha, N. L., Elbeltagi, A., Mehan, S., Malik, A., & Yousuf, A. (2022). Comparative study on morphometric analysis and RUSLE-based approaches for micro-watershed prioritization using remote sensing and GIS. *Arabian Journal of Geosciences*, *15*(7), 564. doi:10.100712517-022-09837-2

Kushwaha, N. L., Rajput, J., Elbeltagi, A., Elnaggar, A. Y., Sena, D. R., Vishwakarma, D. K., Mani, I., & Hussein, E. E. (2021). Data intelligence model and meta-heuristic algorithms-based pan evaporation modelling in two different agro-climatic zones: A case study from Northern India. *Atmosphere (Basel)*, *12*(12), 1654. doi:10.3390/atmos12121654

Kussul, N., Lavreniuk, M., Skakun, S., & Shelestov, A. (2017). Deep learning classification of land cover and crop types using remote sensing data. *IEEE Geoscience and Remote Sensing Letters*, *14*(5), 778–782. doi:10.1109/LGRS.2017.2681128

Latifi, S., Hauser, M., Raheli, H., Moghaddam, S. M., Viira, A.-H., Ozuyar, P. G., & Azadi, H. (2020). Impacts of organizational arrangements on conservation agriculture: Insights from interpretive structural modeling in Iran. *Agroecology and Sustainable Food Systems*. doi:10.1080/21683565.2020.1751375

Lee, K., Silva, B. N., & Han, K. (2020). Deep learning entrusted to fog nodes (DLEFN) based smart agriculture. *Applied Sciences (Basel, Switzerland)*, *10*(4), 1544. doi:10.3390/app10041544

Li, L., Zhang, S., & Wang, B. (2021). Plant Disease Detection and Classification by Deep Learning—A Review. *IEEE Access, 9*, 56683–56698.

Li, C., Li, H., Liu, Z., Li, B., & Huang, Y. (2021). SeedSortNet: A rapid and highly effificient lightweight CNN based on visual attention for seed sorting. *PeerJ. Computer Science*, *7*, e639. doi:10.7717/peerj-cs.639 PMID:34435095

Liu, B.-P., Qin, P., Liu, Y.-Y., Yuan, L., Gu, L.-X., & Jia, C.-X. (2018). Mental disorders and suicide attempt in rural China. *Psychiatry Research*, *261*, 190–196. doi:10.1016/j.psychres.2017.12.087 PMID:29309959

Lo Bianco, M., Grillo, O., Cañadas, E., Venora, G., & Bacchetta, G. (2017). Inter-and intraspecific diversity in Cistus L.(Cistaceae) seeds, analysed with computer vision techniques. *Plant Biology*, *19*(2), 183–190. doi:10.1111/plb.12529 PMID:27917577

Lombardi, M., & Fargnoli, M. (2018). Prioritization of hazards by means of a QFD-based procedure. *Safety and Security Studies*, 163.

Luthra, S., Kumar, S., Kharb, R., Ansari, M. F., & Shimm, S. L. (2014). Adoption of smart grid technologies: An analysis of interactions among barriers. *Renewable & Sustainable Energy Reviews*, *33*, 554–565. doi:10.1016/j.rser.2014.02.030

Maduranga, M. W. P., & Abeysekera, R. (2020). Machine learning applications in IoT based agriculture and smart farming: A review. *Int. J. Eng. Appl. Sci. Technol*, *4*(12), 24–27. doi:10.33564/IJEAST.2020.v04i12.004

Magomadov, V. S. (2019, December). Deep learning and its role in smart agriculture. *Journal of Physics: Conference Series*, *1399*(4), 044109. doi:10.1088/1742-6596/1399/4/044109

Mahbub, M. (2020). A smart farming concept based on smart embedded electronics, internet of things and wireless sensor network. *Internet of Things*, *9*, 100161. doi:10.1016/j.iot.2020.100161

Makalew, J., & Kusuma, Z., Sugiyanto, Tamod, Z. (2013). The Influence of Farmer Characteristic and Farming to the Farmer Motivation on Using Organic Manure (Case Study in East Tombatu District, South-East Minahasa Regency). *International Journal of Engineering Inventions*, *3*(1), 43–51.

Malekirad, A. A., Faghih, M., Mirabdollahi, M., Kiani, M., Fathi, A., & Abdollahi, M. (2013). Neurocognitive, mental health, and glucose disorders in farmers exposed to Organophosphorus pesticides. *(Report). Arhiv za Higijenu Rada i Toksikologiju*, *64*(1), 1–8. doi:10.2478/10004-1254-64-2013-2296 PMID:23705196

Mandal, A., & Deshmukh, S. G. (1994). Vendor selection using interpretive structural modeling (ISM). *International Journal of Operations & Production Management*, *14*(6), 52–59. doi:10.1108/01443579410062086

Mardle, S. J., Pascoe, S., & Tamiz, M. (2000). An investigation of genetic algorithms for the optimization of multi-objective fisheries bioeconomic models. *International Transactions in Operational Research*, *7*(1), 33–49. doi:10.1111/j.1475-3995.2000.tb00183.x

Marko, R. S., Awasthi, A., Jehan, M., & Bose, A. (2018). A Correlation Analysis for Estimation of Height from Head Breadth and Vertex Height. *IOSR Journal of Dental and Medical Sciences*, *17*(4), 87–91.

Marotz-Baden, R., & Mattheis, C. (1994). Daughters-in-law and stress in two generation farm families. *Family Relations*, *43*(2), 132–137. doi:10.2307/585314

Mathiyazhagan, K., Govindan, K., NoorulHaq, A., & Geng, Y. (2013). An ISM approach for the barrier analysis in implementing green supply chain management. *Journal of Cleaner Production*, *47*, 283–297. doi:10.1016/j.jclepro.2012.10.042

McCall, J. (2005). Genetic algorithms for modelling and optimisation. *Journal of Computational and Applied Mathematics*, *184*(1), 205–222. doi:10.1016/j.cam.2004.07.034

McCurdy, S. A., & Carroll, D. J. (2000). Agricultural injury. *American Journal of Industrial Medicine*, *38*(4), 463–480. doi:10.1002/1097-0274(200010)38:4<463::AID-AJIM13>3.0.CO;2-N PMID:10982988

McGlym, S., & Walters, D. (2019). Agricultural Robots: Future Trends for Autonomous Farming. *Journal of Emerging Technologies and Innovative Research, 6*(4).

McInnes, C. R. (1999). Solar radiation pressure. In *Solar sailing* (pp. 32–55). Springer. doi:10.1007/978-1-4471-3992-8_2

McShane, C. J., Quirk, F., & Swinbourne, A. (2016). Development and validation of a work stressor scale for Australian farming families. *The Australian Journal of Rural Health*, *24*(4), 238–245. doi:10.1111/ajr.12261 PMID:26689163

Meeker, B. J., Carruth, A., & Holland, C. B. (2002). Health Hazards and Preventive Measures of Farm Women. *AAOHN Journal*, *50*(7), 307–314. doi:10.1177/216507990205000708 PMID:12141093

Mekonnen, Y., Namuduri, S., Burton, L., Sarwat, A., & Bhansali, S. (2019). Machine learning techniques in wireless sensor network based precision agriculture. *Journal of the Electrochemical Society*, *167*(3), 037522. doi:10.1149/2.0222003JES

Merriott, D. (2016). Factors associated with the farmer suicide crisis in India. *Journal of Epidemiology and Global Health*, *6*(4), 217–227. doi:10.1016/j.jegh.2016.03.003 PMID:27080191

Militante, S. V., Gerardo, B. D., & Dionisio, N. V. (2019). Plant Leaf Detection and Disease Recognition using Deep Learning. *Proceeding of the IEEE Eurasia Conference on IOT, Communication and Engineering (ECICE)*. 10.1109/ECICE47484.2019.8942686

Mishra, D., & Satapathy, S. (2020). An Integrated SWARA, QFD and ISM approach for Agricultural injuries in India. *International Journal of Decision Support System Technology*, *12*(2), 1–24. doi:10.4018/IJDSST.2020040101

Mitra, S., & Kundu, S. (2018). Application of TOPSIS for best domestic refrigerator selection. *International Journal of Research and Analytical Reviews*, *5*(3), 226–231.

Mohanty, B. B. (2005). 'We are like the living dead': Farmer suicides in Maharashtra, western India. *The Journal of Peasant Studies*, *32*(2), 243–276. doi:10.1080/03066150500094485

Mohanty, S. P., Hughes, D. P., & Salathé, M. (2016). Using deep learning for imagebased plant disease detection. *Frontiers in Plant Science*, *7*(September). PMID:27713752

Mohanty, S. P., Hughes, D. P., & Salathé, M. (2016). Using Deep Learning for Image-Based Plant Disease Detection. *Frontiers in Plant Science*, *7*, 1419. doi:10.3389/fpls.2016.01419 PMID:27713752

Mohsin Kabir, M., Quwsar Ohi, A., & Mridha, M. F. A Multi-plant disease diagnosis method using convolutional neural network. arXiv 2020, arXiv:2011.05151.

Moore, J., Chase, J. S., & Ranganathan, P. (2006). *Weatherman: automated, online and predictive thermal mapping and management for data centers. In 2006 IEEE international conference on Autonomic Computing*. IEEE.

Mora, D. C., Quandt, S. A., Chen, H., & Arcury, T. A. (2016). Associations of Poor Housing with Mental Health Among North Carolina Latino Migrant Farmworkers. *Journal of Agromedicine*, *21*(4), 327–334. doi:10.1080/1059924X.2016.1211053 PMID:27409300

Morais, R., Silva, N., Mendes, J., Adão, T., Pádua, L., López-Riquelme, J. A., Pavón-Pulido, N., Sousa, J. J., & Peres, E. (2019). Mysense: A comprehensive data management environment to improve precision agriculture practices. *Computers and Electronics in Agriculture*, *162*, 882–894. doi:10.1016/j.compag.2019.05.028

Morgan, M. I., Hine, D. W., Bhullar, N., Dunstan, D. A., & Bartik, W. (2016). Fracked: Coal seam gas extraction and farmers mental health. *Journal of Environmental Psychology*, *47*, 22–32. doi:10.1016/j.jenvp.2016.04.012

Mortensen, A. K., Dyrmann, M., Karstoft, H., Jørgensen, R. N., & Gislum, R. (2016, June). Semantic segmentation of mixed crops using deep convolutional neural network. In CIGR-AgEng conference (pp. 26-29).

Muangprathub, J., Boonnam, N., Kajornkasirat, S., Lekbangpong, N., Wanichsombat, A., & Nillaor, P. (2019). IoT and agriculture data analysis for smart farm. *Computers and Electronics in Agriculture*, *156*, 467–474. doi:10.1016/j.compag.2018.12.011

Murugamani, C., Shitharth, S., Hemalatha, S., Kshirsagar, P. R., Riyazuddin, K., Naveed, Q. N., Islam, S., Mazher Ali, S. P., & Batu, A. (2022). Machine learning technique for precision agriculture applications in 5G-based internet of things. *Wireless Communications and Mobile Computing*, *2022*, 2022. doi:10.1155/2022/6534238

Muruganantham, P., Wibowo, S., Grandhi, S., Samrat, N. H., & Islam, N. (2022). A systematic literature review on crop yield prediction with deep learning and remote sensing. *Remote Sensing (Basel)*, *14*(9), 1990. doi:10.3390/rs14091990

Muthurasu, S. & Divya. V. (2019). Design and Implementation of Agrobot by Using IOT. *International Journal of Advance Research, Ideas and Innovations in Technology, 5*(2).

Nahatkar, S., Gaur, A., & Pattewa, T. (n.d.). Design of a Home Embedded Surveillance System with Piezoelectric Infrared Sensor & Ultra-Low Alert Power. *International Journal of Advanced Research in Electronics and Communication Engineering, 1*(3).

Nath, B. (2019). Assamese Traditional Icon "Japi": History And Diversity. *RESEARCH REVIEW International Journal of Multidisciplinary*, *4*(8), 29–33.

Newlin Shebiah, R., Deeksha, B., & Aparna, S. (2017). Early Warning System From Threat Of Wild Animals Using Raspberry Pi. *SSRG International Journal of Electronics and Communication Engineering*, 132-136.

Nischitha, K., Vishwakarma, D., Ashwini, M. N., & Manjuraju, M. R. (2020). Crop prediction using machine learning approaches. *International Journal of Engineering Research & Technology (Ahmedabad)*, *9*(08), 23–26.

Nunoo-Mensah, H. (2022). A Survey of Deep Learning Techniques for Maize Leaf Disease Detection: Trends from 2016 to 2021 and Future Perspectives. *Journal of Electrical and Computer Engineering Innovations*, *10*(2), 381–392.

Othman, A., & Owen, L. (2001). The multi dimensionality of Carter Model to measure customer service quality (SQ) in Islamic banking industry: A study in Kuwait finance house. *International Journal of Islamic Financial Services*, *3*(4), 1–12.

Padhy, C., & Raju, P. S. (2020). Mental Health of Farmers–Need of the Hour. *International Journal of Agriculture Environment and Biotechnology*, *13*(1), 7–91. doi:10.30954/0974-1712.1.2020.11

Padula, R. & Head, W. (2003). *Fencing System.* Project Report published by university of Minnesota.

Palanivel, K., & Surianarayanan, C. (2019). An approach for prediction of crop yield using machine learning and big data techniques. *International Journal of Computer Engineering and Technology, 10*(3), 110–118. doi:10.34218/IJCET.10.3.2019.013

Pande, S., & Savenije, H. H. (2016). A sociohydrological model for smallholder farmers in Maharashtra, India. *Water Resources Research, 52*(3), 1923–1947. doi:10.1002/2015WR017841

Pant, J., Pant, R. P., Singh, M. K., Singh, D. P., & Pant, H. (2021). Analysis of agricultural crop yield prediction using statistical techniques of machine learning. *Materials Today: Proceedings, 46*, 10922–10926. doi:10.1016/j.matpr.2021.01.948

Paustian, M., & Theuvsen, L. (2017). Adoption of precision agriculture technologies by German crop farmers. *Precision Agriculture, 18*(5), 701–716. doi:10.100711119-016-9482-5

Peng, C., Wu, G., Liao, T. W., & Wang, H. (2019). Research on multi-agent genetic algorithm based on tabu search for the job shop scheduling problem. *PLoS One, 14*(9), e0223182. doi:10.1371/journal.pone.0223182 PMID:31560722

PlantVillage. (n.d.). *About.* Plant Village. https://plantvillage.psu.edu/

Prajith, A. S. & Nowfiya B.S. (2020). Automatic Agricultural Robot-Agrobot. 2020 IEEE Bangalore Humanitarian Technology Conference. IEEE.

Price, L., & Evans, N. (2009). From stress to distress: Conceptualizing the British family farming patriarchal way of life. *Journal of Rural Studies, 25*(1), 1–11. doi:10.1016/j.jrurstud.2008.03.008

Priyadharsini, M. & Indhumathi, U. (2019). Automatic Irrigation System Using Soil. *International Journal of Engineering Trends and Technology, 67*(3).

Putzu, L., Di Ruberto, C., & Fenu, G. (2016). A mobile application for leaf detection in complex background using saliency maps. *Advanced Concepts for Intelligent Vision Systems: 17th International Conference, ACIVS 2016, Lecce, Italy, October 24-27, 2016 Proceedings, 17*, 570–581.

Qiao, Y., Su, D., Kong, H., Sukkarieh, S., Lomax, S., & Clark, C. (2019). Individual cattle identification using a deep learning based framework. *IFAC-PapersOnLine, 52*(30), 318–323. doi:10.1016/j.ifacol.2019.12.558

Qiu, Z., Chen, J., Zhao, Y., Zhu, S., He, Y., & Zhang, C. (2018). Variety identification of single rice seed using hyperspectral imaging combined with convolutional neural network. *Applied Sciences (Basel, Switzerland), 8*(2), 212. doi:10.3390/app8020212

Rahnemoonfar, M., Dobbs, D., Yari, M., & Starek, M. J. (2019). DisCountNet: Discriminating and counting network for real-time counting and localization of sparse objects in high-resolution UAV imagery. *Remote Sensing (Basel), 11*(9), 1128. doi:10.3390/rs11091128

Rahnemoonfar, M., & Sheppard, C. (2017). Deep count: Fruit counting based on deep simulated learning. *Sensors (Basel), 17*(4), 905. doi:10.339017040905 PMID:28425947

Rajak, R. K., Pawar, A., Pendke, M., Shinde, P., Rathod, S., & Devare, A. (2017). Crop recommendation system to maximize crop yield using machine learning technique. *International Research Journal of Engineering and Technology*, *4*(12), 950–953.

Raj, T., Shankar, R., & Suhaib, M. (2008). An ISM approach for modeling the enablers of flexible manufacturing system: The case for India. *International Journal of Production Research*, *46*(24), 6883–6912. doi:10.1080/00207540701429926

Ranjitha, B., & Nikhitha, M. N. (2019). Solar Powered Autonomous Multipurpose Agricultural Robot Using Bluetooth/Android App. In *3rd International Conference on Electronics, Communication and Aerospace Technology*. IEEE.

Rao, M. S., Singh, A., Reddy, N. S., & Acharya, D. U. (2022). Crop prediction using machine learning. *Journal of Physics: Conference Series*, *2161*(1), 012033. doi:10.1088/1742-6596/234/1/012033

Ratnadip, S. & Patil, S. (2021). Design and Development of Multipurpose Solar Powered Agro-Robot. *International Journal of Research in Engineering and Science*, *9*(8).

Ravi, V., & Shankar, R. (2005). Analysis of interactions among the barriers of reverse logistics. *Technological Forecasting and Social Change*, *72*(8), 1011–1029. doi:10.1016/j.techfore.2004.07.002

Rayens, M. K., & Reed, D. B. (2014). Predictors of Depressive Symptoms in Older Rural Couples: The Impact of Work, Stress and Health. *The Journal of Rural Health*, *30*(1), 59–68. doi:10.1111/jrh.12028 PMID:24383485

Reddy, D. J., & Kumar, M. R. (2021, May). Crop yield prediction using machine learning algorithm. In *2021 5th International Conference on Intelligent Computing and Control Systems (ICICCS)* (pp. 1466-1470). IEEE. 10.1109/ICICCS51141.2021.9432236

Ren, C., Kim, D. K., & Jeong, D. (2020). A survey of deep learning in agriculture: Techniques and their applications. *Journal of Information Processing Systems*, *16*, 1015–1033.

Rezk, N. G., Hemdan, E. E. D., Attia, A. F., El-Sayed, A., & El-Rashidy, M. A. (2021). An efficient IoT based smart farming system using machine learning algorithms. *Multimedia Tools and Applications*, *80*(1), 773–797. doi:10.100711042-020-09740-6

Riese, F. M., & Keller, S. (2019). Soil texture classification with 1D convolutional neural networks based on hyperspectral data. *arXiv preprint arXiv:1901.04846*.

Rivas, A., Chamoso, P., González-Briones, A., & Corchado, J. M. (2018). Detection of cattle using drones and convolutional neural networks. *Sensors (Basel)*, *18*(7), 2048. doi:10.339018072048 PMID:29954080

Rony, M., Barai, D., & Hasan, Z. (2021, July). Cattle external disease classification using Deep Learning Techniques. In *2021 12th International Conference on Computing Communication and Networking Technologies (ICCCNT)* (pp. 1-7). IEEE. 10.1109/ICCCNT51525.2021.9579662

Rostamabadi, A., Jahangiri, M., Naderi Mansourabadi, B., Javid, M., Ghorbani, M., & Banaee, S. (2019). Prevalence of chronic diseases and occupational injuries and their influence on the health-related quality of life among farmers working in small-farm enterprises. *Journal of Agromedicine*, *24*(3), 248–256. doi:10.1080/1059924X.2019.1592047 PMID:30880615

Roy, P., Tremblay, G., Oliffe, J. L., Jbilou, J., & Robertson, S. (2013). Male farmers with mental health disorders: A scoping review. *The Australian Journal of Rural Health*, *21*(1), 3–7. doi:10.1111/ajr.12008 PMID:23384130

Saeed, F., Khan, M. A., Sharif, M., Mittal, M., Goyal, L. M., & Roy, S. (2021). Deep neural network features fusion and selection based on PLS regression with an application for crops diseases classification. *Applied Soft Computing*, *103*, 107164. doi:10.1016/j.asoc.2021.107164

Saeidi, S. (2018). Solving the Rubik's Cube using Simulated Annealing and Genetic Algorithm. *International Journal of Education and Management Engineering*, *8*(1), 1–10. doi:10.5815/ijeme.2018.01.01

Safety and health in agriculture (2011). *ILO code of practice.* International Labour Office - Geneva: ILO.

Sage, A. (1977). *Interpretive structural modeling: Methodology for large-scale systems*. McGraw-Hill.

Sa, I., Ge, Z., Dayoub, F., Upcroft, B., Perez, T., & McCool, C. (2016). Deepfruits: A fruit detection system using deep neural networks. *Sensors (Basel)*, *16*(8), 1222. doi:10.339016081222 PMID:27527168

Sa, I., Popović, M., Khanna, R., Chen, Z., Lottes, P., Liebisch, F., Nieto, J., Stachniss, C., Walter, A., & Siegwart, R. (2018). WeedMap: A large-scale semantic weed mapping framework using aerial multispectral imaging and deep neural network for precision farming. *Remote Sensing (Basel)*, *10*(9), 1423. doi:10.3390/rs10091423

Saikia, S.B. (2012). *Design Audit Report.* Mukhkushi Japi making Cluster, 2012.

Sami, M., Khan, S. Q., Khurram, M., Farooq, M. U., Anjum, R., Aziz, S., Qureshi, R., & Sadak, F. (2022). A deep learning-based sensor modeling for smart irrigation system. *Agronomy (Basel)*, *12*(1), 212. doi:10.3390/agronomy12010212

Sangeeta, S. G. (2020). Design and implementation of crop yield prediction model in agriculture. *International Journal of Scientific & Technology Research*, *8*(1), 544–549.

Santos, L. (2020). Deep Learning Applications in Agriculture: A Short Review. *2019 Fourth Iberian Robotics Conference*. 10.1007/978-3-030-35990-4_12

Santos, L., Santos, F. N., Oliveira, P. M., & Shinde, P. (2020). Deep learning applications in agriculture: A short review. *Robot 2019: Fourth Iberian Robotics Conference Advanced Robotics*, *1*, 139–151.

Santos, T. T., de Souza, L. L., dos Santos, A. A., & Avila, S. (2020). Grape detection, segmentation, and tracking using deep neural networks and three-dimensional association. *Computers and Electronics in Agriculture, 170*, 105247. doi:10.1016/j.compag.2020.105247

Saravanan, N. (2021). Agricultural Robot using Machine Learning and Computer Vision. *International Research Journal of Engineering and Technology, 8*(3).

Sari, R. M., Rizkya, I., Syahputri, K., & Siregar, I. (2018). Alternative of raw material's suppliers using TOPSIS method in chicken slaughterhouse industry. *IOP Conference Series. Materials Science and Engineering, 309*(1), 012041. doi:10.1088/1757-899X/309/1/012041

Satapathy, S. (2014). ANN, QFD and ISM approach for framing electricity utility service in India for consumer satisfaction. *International Journal of Services and Operations Management, 18*(4), 404–428. doi:10.1504/IJSOM.2014.063243

Sau, S., Ucchesu, M., D'hallewin, G., & Bacchetta, G. (2019). Potential use of seed morpho-colourimetric analysis for Sardinian apple cultivar characterisation. *Computers and Electronics in Agriculture, 162*, 373–379. doi:10.1016/j.compag.2019.04.027

Sau, S., Ucchesu, M., Dondini, L., De Franceschi, P., D'hallewin, G., & Bacchetta, G. (2018). Seed morphometry is suitable for apple-germplasm diversity-analyses. *Computers and Electronics in Agriculture, 151*, 118–125. doi:10.1016/j.compag.2018.06.002

Scarth, R. D., Stallones, L., Zwerling, C., & Burmeister, L. F. (2000). The prevalence of depressive symptoms and risk factors among Iowa and Colorado farmers. *American Journal of Industrial Medicine, 37*(4), 382–389. doi:10.1002/(SICI)1097-0274(200004)37:4<382::AID-AJIM8>3.0.CO;2-4 PMID:10706750

Scheyett, A., Bayakly, R., & Whitaker, M. (2019). Characteristics and contextual stressors in farmer and agricultural worker suicides in Georgia from 2008–2015. *Rural Mental Health, 43*(2-3), 61–72. doi:10.1037/rmh0000114

Schmidhuber, J. (2015). Deep learning in neural networks: An overview. *Neural Networks, 61*, 85–117. doi:10.1016/j.neunet.2014.09.003 PMID:25462637

Sekulic, M., Pejic, V., Brezocnik, M., Gostimirović, M., & Hadzistevic, M. (2018). Prediction of surface roughness in the ball-end milling process using response surface methodology, genetic algorithms, and grey wolf optimizer algorithm. *Advances in Production Engineering & Management, 13*(1), 18–30. doi:10.14743/apem2018.1.270

Shah, J. P., Prajapati, H. B., & Dabhi, V. K. (2016). A survey on detection and classification of rice plant diseases. In *2016 IEEE Int. Conf. Curr. Trends Adv. Comput.* IEEE. 10.1109/ICCTAC.2016.7567333

Shamshiri, S. & Weltzien, C. (2018). Research and Development in Agricultural Robotics: A Perspective of Digital Farming. *International Journal of Agricultural and Biological Engineering, 11*(4).

Shanmugasundaram, R., & Pavithra, S. (2017). IoT based animal tracking and monitoring system in Zoo. *South Asian Journal of Engineering and Technology, 3*(2).

Sharma, S. & Shah, D. (2016). A Practical Animal Detection and Collision Avoidance System Using Computer Vision Technique. *Special Section on innovations in electrical and computer engineering education.*

Sharma, A., Jain, A., Gupta, P., & Chowdary, V. (2020). Machine learning applications for precision agriculture: A comprehensive review. *IEEE Access : Practical Innovations, Open Solutions, 9*, 4843–4873. doi:10.1109/ACCESS.2020.3048415

Sharma, H., Haque, A., & Jaffery, Z. A. (2019). Smart agriculture monitoring using energy harvesting Internet of Things (EH-IoT). *World Scientific News*, (121), 17–21.

Sharma, R., Kamble, S. S., Gunasekaran, A., Kumar, V., & Kumar, A. (2020). A systematic literature review on machine learning applications for sustainable agriculture supply chain performance. *Computers & Operations Research, 119*, 104926. doi:10.1016/j.cor.2020.104926

Sharma, S. U., & Shah, D. J. (2017). A Practical Animal Detection and Collision Avoidance System Using Computer Vision Technique. *IEEE Access : Practical Innovations, Open Solutions, 5*, 347–358. doi:10.1109/ACCESS.2016.2642981

Shen, L., Song, X., Wu, Y., Liao, S., & Zhang, X. (2016). Interpretive structural modeling based factor analysis on the implementation of emission trading system in the Chinese building sector. *Journal of Cleaner Production, 127*, 214–227. doi:10.1016/j.jclepro.2016.03.151

Shen, W., Hu, H., Dai, B., Wei, X., Sun, J., Jiang, L., & Sun, Y. (2020). Individual identification of dairy cows based on convolutional neural networks. *Multimedia Tools and Applications, 79*(21-22), 14711–14724. doi:10.100711042-019-7344-7

Shruthi, U., Nagaveni, V., & Raghavendra, B. K. (2019). A Review on Machine Learning Classification Techniques for Plant Disease Detection. In *2019 5th Int. Conf. Adv. Comput. Commun. Syst.* (pp. 281–4). IEEE. 10.1109/ICACCS.2019.8728415

Shwetal Ramesh Gund, V. P. (2015, July). A Review on Agricultural Robots. *International Journal of Advanced Research in Computer Engineering and Technology, 4*(7).

Simsek, Z., Ersin, F., & Kirmizitoprak, E. (2015). Development of the Seasonal Migrant Agricultural Worker Stress Scale in Sanliurfa, Southeast Turkey. *Journal of Agromedicine, 21*(1), 56–60. do i:10.1080/1059924X.2015.1106372 PMID:26479204

Sladojevic, S., Arsenovic, M., Anderla, A., Culibrk, D., & Stefanovic, D. (2016). Deep neural networks based recognition of plant diseases by leaf image classification. *Computational Intelligence and Neuroscience, 2016*, 2016. doi:10.1155/2016/3289801 PMID:27418923

Sneha, A., Amrita, E., & Abirami. (2015). Agricultural Robot for Automatic Ploughing and Seeding. *2015 IEEE Technologies Innovation in ICT for Agriculture and Rurla Development.* IEEE.

Stallones, L., & Beseler, C. (2002). Pesticide illness, farm practices, and neurological symptoms among farm residents in Colorado. *Environmental Research*, *90*(2), 89–97. doi:10.1006/enrs.2002.4398 PMID:12483798

Staniford, A. K., Dollard, M. F., & Guerin, B. (2009). Stress and help-seeking for drought-stricken citrus growers in the Riverland of South Australia. *The Australian Journal of Rural Health*, *17*(3), 147–154. doi:10.1111/j.1440-1584.2009.01059.x PMID:19469779

Subeesh, A., Bhole, S., Singh, K., Chandel, N. S., Rajwade, Y. A., Rao, K. V. R., Kumar, S. P., & Jat, D. (2022). Deep convolutional neural network models for weed detection in polyhouse grown bell peppers. *Artificial Intelligence in Agriculture*, *6*, 47–54. doi:10.1016/j.aiia.2022.01.002

Suchithra, M. S., & Pai, M. L. (2020). Improving the prediction accuracy of soil nutrient classification by optimizing extreme learning machine parameters. *Information Processing in Agriculture*, *7*(1), 72–82. doi:10.1016/j.inpa.2019.05.003

Suchithra, M., Asuwini, T., Charumathi, M. C., & Lal, R. N. (2018). Sensor data validation. *International Journal of Pure and Applied Mathematics*, *119*(12), 14327–14335.

Sudarsan, B., Ji, W., Biswas, A., & Adamchuk, V. (2016). Microscope-based computer vision to characterize soil texture and soil organic matter. *Biosystems Engineering*, *152*, 41–50. doi:10.1016/j.biosystemseng.2016.06.006

Sujawat, G. S. (2021). Application of Artificial Intelligence in detection of diseases in plants: A Survey. *Turkish Journal of Computer and Mathematics Education*, *12*(3), 3301–3305.

Sundar, M. (1999). Suicide in farmers in India. *The British Journal of Psychiatry*, *175*(6), 585–586. doi:10.1192/bjp.175.6.585c PMID:10789361

Taheri-Garavand, A., Nasiri, A., Fanourakis, D., Fatahi, S., Omid, M., & Nikoloudakis, N. (2021). Automated in situ seed variety identification via deep learning: A case study in chickpea. *Plants*, *10*(7), 1406. doi:10.3390/plants10071406 PMID:34371609

Tang, J., Wang, D., Zhang, Z., He, L., Xin, J., & Xu, Y. (2017). Weed identification based on K-means feature learning combined with convolutional neural network. *Computers and Electronics in Agriculture*, *135*, 63–70. doi:10.1016/j.compag.2017.01.001

Tejaswani, G. & Afroz, B. (2018). Agriculture Robotics. *International Journal of Engineering and Computer Science, 7(*3).

Terrazas, S. R., & McCormick, A. (2018). Coping Strategies That Mitigate Against Symptoms of Depression Among Latino Farmworkers. *Hispanic Journal of Behavioral Sciences*, *40*(1), 57–72. doi:10.1177/0739986317752923

Torske, M. O., Hilt, B., Bjørngaard, J. H., Glasscock, D., & Krokstad, S. (2015). Disability pension and symptoms of anxiety and depression: A prospective comparison of farmers and other occupational groups. The HUNT Study, Norway. *BMJ Open*, *2015*(11), 5. doi:10.1136/bmjopen-2015-009114 PMID:26525724

Torske, M. O., Hilt, B., Glasscock, D., Lundqvist, P., & Krokstad, S. (2016). Anxiety and depression symptoms among farmers: The HUNT study, Norway. *Journal of Agromedicine*, *21*(1), 24–33. doi:10.1080/1059924X.2015.1106375 PMID:26488439

Tran, T. T., Choi, J. W., Le, T. T. H., & Kim, J. W. (2019). A comparative study of deep CNN in forecasting and classifying the macro-nutrient deficiencies on development of tomato plant. *Applied Sciences (Basel, Switzerland)*, 9.

Trnovszky, T., Kamencay, P., Orjesek, R., Benco, M., & Sykora, P. (2017). *Animal recognition system based on convolutional neural network*. Semantic Scholar.

Ünal, Z. (2020). Smart farming becomes even smarter with deep learning—A bibliographical analysis. *IEEE Access : Practical Innovations, Open Solutions*, *8*, 105587–105609. doi:10.1109/ACCESS.2020.3000175

Usha Kumari, Ch., Jeevan Prasad, S., & Mounika, G. (2019). Leaf Disease Detection: Feature Extraction with K-means clustering and Classification with ANN. *3rd International Conference on Computing Methodologies and Communication (ICCMC)*, Erode, India.

Vaishnav, P., Choudhary, N., & Jain, K. (2017). Traveling salesman problem using genetic algorithm: A survey. *International Journal of Scientific Research in Computer Science. Engineering and Information Technology*, *2*(3), 105–108.

Van Klompenburg, T., Kassahun, A., & Catal, C. (2020). Crop yield prediction using machine learning: A systematic literature review. *Computers and Electronics in Agriculture*, *177*, 105709. doi:10.1016/j.compag.2020.105709

Vanbrabant, Y., Delalieux, S., Tits, L., Pauly, K., Vandermaesen, J., & Somers, B. (2020). Pear flower cluster quantification using RGB drone imagery. *Agronomy (Basel)*, *10*(3), 407. doi:10.3390/agronomy10030407

Vazquez-Carillo. (2020) *Crop Yield Prediction using Machine Learning and Satellite Imagery.*

Villacrés, J. F., & Auat Cheein, F. (2020). Detection and characterization of cherries: A deep learning usability case study in Chile. *Agronomy (Basel)*, *10*(6), 835. doi:10.3390/agronomy10060835

Vishwakarma, D. K., Pandey, K., Kaur, A., Kushwaha, N. L., Kumar, R., Ali, R., Elbeltagi, A., & Kuriqi, A. (2022). Methods to estimate evapotranspiration in humid and subtropical climate conditions. *Agricultural Water Management*, *261*, 107378. doi:10.1016/j.agwat.2021.107378

Walker, J. L., & Walker, L. J. S. (1988). Self-reported stress symptoms in farmers. *Journal of Clinical Psychology*, *44*(1), 10–16. doi:10.1002/1097-4679(198801)44:1<10::AID-JCLP2270440103>3.0.CO;2-6 PMID:3343357

Walker, L. S., & Walker, J. L. (1987). Stressors and symptoms predictive of distress in farmers. *Family Relations*, *36*(4), 374–378. doi:10.2307/584486

Wallelign, S., Polceanu, M., & Buche, C. (2018). Soybean plant disease identification using a convolutional neural network. In: Proc. 31st Int. Florida Artif. Intell. Res. Soc. Conf. FLAIRS, Florida, USA.

Walter, A., Finger, R., Huber, R., & Buchmann, N. (2017). Opinion: Smart farming is key to developing sustainable agriculture. *Proceedings of the National Academy of Sciences of the United States of America*, *114*(24), 6148–6150. doi:10.1073/pnas.1707462114 PMID:28611194

Wang, G. H., Wang, Y., & Zhao, T. (2008). Analysis of interactions among the barriers to energy saving in China. *Energy Policy*, *36*(6), 1879–1889. doi:10.1016/j.enpol.2008.02.006

Wang, S. H., & Chen, Y. (2018). Fruit category classification via an eight-layer convolutional neural network with parametric rectified linear unit and dropout technique. *Multimedia Tools and Applications*, *79*(21-22), 15117–15133. doi:10.100711042-018-6661-6

Warfield, J. W. (1974). Developing interconnected matrices in structural modelling, IEEE transcript on systems. *Men and Cybernetics*, *4*, 51–81.

Wästerlund, D. S. (2018). Managing heat in agricultural work: increasing worker safety and productivity by controlling heat exposure. *Forestry Working Paper No. 1.* Rome, FAO.

Weary, D. M., Huzzey, J. M., & Von Keyserlingk, M. A. G. (2009). Board-invited review: Using behavior to predict and identify ill health in animals. *Journal of Animal Science*, *87*(2), 770–777. doi:10.2527/jas.2008-1297 PMID:18952731

Weigel, R. R., Weigel, D. J., & Blundall, J. (1987). Stress, coping, and satisfaction: Generational differences in farm families. *Family Relations*, *36*(1), 45–48. doi:10.2307/584646

Weisskopf, M. G., Moisan, F., Tzourio, C., Rathouz, P. J., & Elbaz, A. (2013). Pesticide Exposure and Depression Among Agricultural Workers in France. *American Journal of Epidemiology*, *178*(7), 1051–1058. doi:10.1093/aje/kwt089 PMID:23851580

Wilson. (2011). Electric Fence. *Handbook of Texas*. Project report published by the Texas State Historical Association.

Xiao, J., Liu, G., Wang, K., & Si, Y. (2022). Cow identification in free-stall barns based on an improved Mask R-CNN and an SVM. *Computers and Electronics in Agriculture*, *194*, 106738. doi:10.1016/j.compag.2022.106738

Xie, B., & Liu, J. (2021). Research Progress on Autonomous Navigation Technology of Agricultural Robot. *2021 IEEE 11th Annual International Conference on Cyber technology in Automation, Control and Intelligent Systems*. IEEE .

Xie, Z., Singh, A., Uang, J., Narayan, K. S., & Abbeel, P. (2013). Multimodal blending for high-accuracy Instance cognition. In *IEEE/RSJ International Conference on Intelligent Robots and Systems*. IEEE. 10.1109/IROS.2013

Yang, Q., Xiao, D., & Lin, S. (2018). Feeding behavior recognition for group-housed pigs with the Faster R-CNN. *Computers and Electronics in Agriculture*, *155*, 453–460. doi:10.1016/j.compag.2018.11.002

Yang, S., Logan, J., & Coffey, D. L. (1995). Mathematical formulae for calculating the base temperature for growing degree days. *Agricultural and Forest Meteorology*, *74*(1-2), 61–74. doi:10.1016/0168-1923(94)02185-M

Yang, X.-S. (2012). *Flower pollination algorithm for global optimization.* In: *International conference on unconventional computing and natural computation.* Springer, Berlin. 10.1007/978-3-642-32894-7_27

Yazd, S. D., Wheeler, S. A., & Zuo, A. (2019). Key Risk Factors Affecting Farmers' Mental Health: A Systematic Review. *International Journal of Environmental Research and Public Health*, *16*(23), 4849. doi:10.3390/ijerph16234849 PMID:31810320

Yin, H., Xu, G., Tian, H., Yang, G., Wardenaar, K. J., & Schoevers, R. A. (2018). The prevalence, age-of-onset and the correlates of DSM-IV psychiatric disorders in the Tianjin Mental Health Survey (TJMHS). *Psychological Medicine*, *48*(3), 473–487. doi:10.1017/S0033291717001878 PMID:28714421

Zhang, C., Marzougui, A., & Sankaran, S. (2020). High-resolution satellite imagery applications in crop phenotyping: An overview. *Computers and Electronics in Agriculture*, *175*, 105584. doi:10.1016/j.compag.2020.105584

Zhang, C., Yue, P., Di, L., & Wu, Z. (2018). Automatic identification of center pivot irrigation systems from landsat images using convolutional neural networks. *Agriculture*, *8*(10), 147. doi:10.3390/agriculture8100147

Zhao, H., & Kong, F. (2016). Research and Applications of Shop Scheduling Based on Genetic Algorithms. *Brazilian Archives of Biology and Technology*, *59*(SPE.

Zhao, G., Quan, L., Li, H., Feng, H., Li, S., Zhang, S., & Liu, R. (2021). Real-time recognition system of soybean seed full-surface defects based on deep learning. *Computers and Electronics in Agriculture*, *187*, 106230. doi:10.1016/j.compag.2021.106230

Zhao, J., & Qu, J. (2019). A Detection Method for Tomato Fruit Common Physiological Diseases Based on YOLOv2. *Proceeding of the 10th International Conference on Information Technology in Medicine and Education*. 10.1109/ITME.2019.00132

Zhao, Y., Liu, L., Xie, C., Wang, R., Wang, F., Bu, Y., & Zhang, S. (2020). An effective automatic system deployed in agricultural Internet of Things using Multi-Context Fusion Network towards crop disease recognition in the wild. *Applied Soft Computing*, *89*, 106128. doi:10.1016/j.asoc.2020.106128

Zheng, Y. Y., Kong, J. L., Jin, X. B., Wang, X. Y., Su, T. L., & Zuo, M. (2019). CropDeep: The crop vision dataset for deep-learning-based classification and detection in precision agriculture. *Sensors (Basel)*, *19*(5), 1058. doi:10.339019051058 PMID:30832283

Zhou, P., Fan, L. W., & Zhou, D. Q. (2010). Data aggregation in constructing composite indicators: A perspective of information loss. *Expert Systems with Applications*, *37*(1), 360–365. doi:10.1016/j.eswa.2009.05.039

Zhu, N., Liu, X., Liu, Z., Hu, K., Wang, Y., Tan, J., Huang, M., Zhu, Q., Ji, X., Jiang, Y., & Guo, Y. (2018). Deep learning for smart agriculture: Concepts, tools, applications, and opportunities. *International Journal of Agricultural and Biological Engineering*, *11*(4), 32–44. doi:10.25165/j.ijabe.20181104.4475

Zhu, S., Zhou, L., Gao, P., Bao, Y., He, Y., & Feng, L. (2019). Near-infrared hyperspectral imaging combined with deep learning to identify cotton seed varieties. *Molecules (Basel, Switzerland)*, *24*(18), 3268. doi:10.3390/molecules24183268 PMID:31500333

Zole, K., & Gedam, S. (2018). Agricultural Robot. *International Research Journal of Engineering and Technology, 5*(2).

Zulqarnain, M., & Dayan, F. (2017). Choose Best Criteria for Decision Making Via Fuzzy Topsis Method. *Mathematics and Computer Science*, 2(6), 113–119. doi:10.11648/j.mcs.20170206.14

About the Contributors

Mohammad Farukh Hashmi (Senior Member, IEEE) received the B.E. degree in electronics and communication engineering from MIT Mandsaur/RGPV Bhopal University in 2007, the M.E. degree in digital techniques and instrumentation from SGSITS Indore/RGPV Bhopal University, in 2010. Dr. Hashmi Received Ph.D. degree from the Visvesvaraya National Institute of Technology (VNIT), Nagpur in 2015, under the supervision of Dr. Avinash G. Keskar. He is currently an Assistant Professor with the Department of Electronics and Communication Engineering, National Institute of Technology, Warangal. He has published up to 75 articles in International/National Journals/Conferences. He has a teaching and research experience of 12 years. His current research interests include computer vision, machine vision, machine learning, deep learning, embedded systems, Internet of things, digital signal processing, image processing, and digital IC design. He is a senior member of IEEE, Life member of IETE, Life member ISTE, and Life member of IAENG societies.

Avinash G. Keskar (Senior Member, IEEE) was born in Nagpur, India, in 1959. He received the B.E. degree (Hons.) from the Visvesvaraya National Institute of Technology (VNIT), Nagpur, in 1979, the M.E. degree (Hons.) from the Indian Institute of Science (IISc), Bangalore, in 1983, and the Ph.D. degree from Nagpur University, in 1997. He has 30 years of teaching experience and seven years of industrial experience. He is currently working as a Professor and the Head of the Department of Electronics and Communication Engineering, VNIT. His current research interests include computer vision, soft computing, embedded systems, and fuzzy logic. He is a Senior Member of FIETE, FIE, and LMISTE.

Bhuvaneswari Amma N.G. obtained her degree in Information Technology with Distinction in 2004 from Manonmaniam Sundaranar University, Tirunelveli, Tamil Nadu and Postgraduate degree in Computer Science and Engineering in 2009 from College of Engineering, Anna University Guindy Campus, Chennai, Tamil Nadu. She was awarded Ph.D. degree by the National Institute of Technology Tiruchirappalli, Tamil Nadu in 2020. Since 2004, she has been in the teaching profession and currently, she is an Assistant Professor (Senior) in the School of Computer Science and Engineering, Vellore Institute of Technology (VIT), Chennai, Tamil Nadu. She has 11 years of teaching experience and 7 years of research experience. Her research interests include data science and cyber security. She received awards such as, Best Paper Award, Best Poster Award, and CSIR Travel Grant. She has visited Nanyang Technological University Singapore and South Korea.

Devershi Pallavi Bhatt is Professor & Head Department of Computer Applications School of Basic Sciences Manipal University Jaipur

Shivnath Ghosh is a professor and Head of the Computer science and engineering department of Brainware University

Muktha Patil is an Assistant professor at Tontadarya College of Engineering Gadag.

Linesh Raja is Associate Professor Department of Computer Applications School of Basic Sciences.

R. Felshiya Rajakumari is an Assistant Professor in the Department of Electrical and Electronics Engineering, Karpagam Institute of Technology, Coimbatore. I have received Master Degree in Power Electronics and Drives in the year 2018 and Bachelor Degree in Electrical and Electronics Engineering in the year 2016. Now I am pursuing Ph.D in the domain Power Electronics. I have entered into final stage to finish my Ph.D which will be completed this year. I have received Gold Medal Award in Master Degree, SSN College of Engineering for "First Rank" in the Anna University Examinations. I have 4.5 years of teaching experience in Karpagam Institute of Technology, Coimbatore. I have received favourite teacher award and Best Young Faculty Award from Novel Researcher Academy. My areas of interest are Power Electronics and Drives, Renewable Energy, Measurements and Instrumentation, Biomedical Instrumentation and Semiconductor Materials. I have published 12 papers in reputed journal, 20 papers in international conferences and 4 papers in book chapter publications and filed 3 patents. I have organized more than 25 seminar/workshop/guest lecturer in technical field. I have attended more than 20 faculty development program handled session on entrepreneur program.. I have received elite certificate for secured in NPTEL exam. Acted as IIC Co-ordinator, ICT Academy Coordinator, Entrepreneurship Co-ordinator, Association In-charges, Exam cell Coordinator, Conference Coordinator, IIT Spoken Tutorial Inviligilator, IIT Spoken Tutorial Coordinator, Graduation Day Program Co-ordinator, Symposium (TEKSPARKS) Co-ordinator, Alumni Coordinator. I have applied 3 projects in TamilNadu State Council for Science and Technology (TNSCST) and one project in Institute for Engineering Research and Publication and received the certificate in first round. I am an active member in IAENG, IFERP, IRED, ENSAV Green Club.

A. Ramalingam completed his MCA and Ph.D (CSE) at Puducherry Technological University formerly called as Pondicherry Engineering College, Puducherry, M.Tech(IT) at AAI-DU, Allahabad and M.Phil(CS) at Periyar University, Salem. He has been working as a Professor and Head in the Department of Computer Applications (MCA Programme), Sri Manakula Vinayagar Engineering College (Autonomous), Puducherry since 2002. He has published more than 25 papers in International Journals, 30 papers in International/National conferences, 2 International and 5 National patents, authored and co-authored for 6 books, 2 book chapters and Invited speakers for reputed institutions. He is a Life member in the professional association of ISTE and CSI.

Piyal Roy is an Assistant Professor in the Department of Computer Science & Engineering at Brainware University, Kolkata, India.

Brijendra Singh received a Ph.D from VIT University, Vellore, Tamilnadu, India. He is an Assistant Professor (Senior) in the School of Information Technology and Engineering at VIT University Vellore, Tamilnadu, India. His research interests include Health Informatics, Soft Computing Techniques, Artificial Intelligence, Rough Set, Knowledge Representations.

S. Gandhimathi Alias Usha is an Associate Professor, with the Department of Electronics and Communication Engineering, Velammal College of Engineering and Technology, Madurai, India. She received her Bachelor's degree in Electronics and Communication Engineering from Madurai Kamaraj University in 2001 and her Master's degree fromA.C.College of Engineering and Technology, Karaikudi, Tamilnadu, India in 2009 and received Ph.D degree in ImageProcessing at Anna University, Chennai in 2019. She has published over 45 Technical papers in International Journals, International/ National Conferences. Her current research includes Image Processing and Remote Sensing. She is a Life member of Indian Society for Technical Education (ISTE) and Indian Society of Remote Sensing (ISRS).

Index

A

Agriculture 1-11, 17, 20-23, 25-29, 31, 33-38, 40-43, 46, 49-61, 65, 87-95, 105-107, 109, 113-114, 122-128, 139-141, 143, 147-148, 150-151, 167, 169-170, 173-174, 178, 190, 195, 214-215, 223-224
Agriculture Productivity 58
Artificial Intelligence Tomato Fruit 93
Aurdino 173
Automated Detection 58
Automation 2, 24-28, 32-33, 35, 43, 96, 104, 106, 110, 190, 216

C

Classification 5-6, 8, 12, 14-15, 19, 23, 34, 36-37, 40-41, 45-46, 49-53, 55-56, 58-60, 64, 66, 68-71, 74-75, 82-83, 85-92, 98, 102, 105, 111, 116, 118, 121, 123, 125, 128, 130-132, 137-139, 179, 186, 192-194, 196-197, 213-216, 219, 223-224
Computer Vision Techniques 1, 44, 47-48, 53, 58, 60, 65, 67-68, 91, 178
Constitutional Neural Network 173
Convolution Neural Network 213, 216, 219
Counting 6, 47, 50, 54, 93-96, 99-105
Crop Management 5, 34, 37, 42, 59, 87, 99, 104, 106-110, 115, 122, 126
Crop Prediction 49, 124-126, 128, 139-141

D

Decision Tree 38, 125, 131, 137-138
Deep Convolutional Neural Networks 2, 17-19, 53, 105, 193-194
Deep Learning 1-4, 6-10, 12, 14, 16-18, 20-28, 32-34, 36, 39, 41-42, 44-45, 48-57, 59-60, 64, 70, 72, 78, 81, 87-91, 96, 101, 104-106, 111-115, 117, 122-123, 127, 139-140, 192-195, 213, 216, 223-224
Deep Learning Framework 1-3, 17
Dense NET 106, 116, 118, 198, 200

E

Efficient Net 199
Ensemble Technique 60, 124-126

F

Farmers 3, 8-10, 25, 30, 32, 35, 37-38, 40-42, 47, 54, 58-60, 87-88, 94-95, 97, 104, 106-110, 113-115, 122, 124-126, 142-146, 148-150, 154, 160-167, 169-174, 179-180, 192-193, 195, 215, 223
Feature Fusion 193, 198-199
Fuzzy-TOPSIS 142, 144, 154, 161-162

G

GA 142, 151-152, 154, 160

H

Harvesting 6, 26-27, 29, 37, 41, 55, 94, 110, 124, 145-146, 150
Head-cap 142

I

Image Augmentation 218-219
Image Processing 1, 37, 39, 44, 46, 52, 90, 93, 186, 188-189, 215-216
India 1, 24, 34, 53, 58, 93-94, 106, 124, 129, 143-145, 148-150, 161-165, 168-171, 173, 190, 192-193, 214-215, 224

K

KNN 72-73, 75, 132, 215, 224

L

Leaf Disease 40, 74, 84, 89-92, 192, 194, 213, 215, 223-224
Livestock Management 41, 47-48
Logistic Regression 125, 132-133, 137-138

M

Machine Learning 2, 6-7, 12, 22-26, 28, 30, 32-38, 40-44, 48-55, 58-59, 65, 67-68, 80, 90-93, 96, 100, 111, 122-130, 132, 134, 139-141, 188, 193, 195, 215, 223-224
Machine Learning Algorithms 6, 35, 37, 52, 55, 58, 67-68, 90, 93, 96, 100, 122, 126-127, 129, 139, 195

N

Naive Bayes 36, 40, 130-131

P

Plant Disease Prediction 20
Plant Diseases 2, 21, 38, 58-61, 64-67, 69, 74, 83-85, 87-88, 90, 115-116, 123, 192-195, 212-216, 223-224
Plant Growth 10
Prediction 6-7, 9-10, 20-21, 34-36, 40-41, 49-52, 55-56, 75, 99, 111, 124-128, 130-132, 134-136, 139-141, 151, 170, 222-223
Pre-Trained Networks 71

R

Random Forest 7, 37, 46, 50, 100, 125-126, 131-132, 137-138
Robot 22, 25-27, 29-33

S

SFO 152, 154, 160
Smart Agriculture 1-2, 5-7, 9-11, 17, 20-23, 34, 42, 55, 105, 121 125, 128, 195
Soil Management 5, 46-47
Sustainability 4, 6, 8-9, 17, 107-110, 165

T

Transfer Learning 16, 19, 60, 64, 67, 72-74, 80-81, 87-88, 90-91, 112, 114, 118-119, 123, 194, 196, 213-216, 223

U

Ultrasonic Sensor 26-27

W

Water Management 41, 45, 48, 56, 107

www.ingramcontent.com/pod-product-compliance
Lightning Source LLC
LaVergne TN
LVHW061537290926
849162LV00004B/825